Material Culture and Technology in Everyday Life

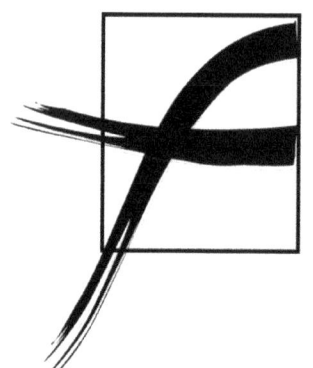

Intersections in Communications and Culture

Global Approaches and Transdisciplinary Perspectives

Cameron McCarthy and Angharad N. Valdivia
General Editors

Vol. 25

PETER LANG
New York • Washington, D.C./Baltimore • Bern
Frankfurt am Main • Berlin • Brussels • Vienna • Oxford

Material Culture and Technology in Everyday Life

ETHNOGRAPHIC APPROACHES

Edited by Phillip Vannini

PETER LANG
New York • Washington, D.C./Baltimore • Bern
Frankfurt am Main • Berlin • Brussels • Vienna • Oxford

Library of Congress Cataloging-in-Publication Data

Material culture and technology in everyday life:
ethnographic approaches / edited by Phillip Vannini.
p. cm. — (Intersections in communications and culture; v. 25)
Includes bibliographical references and index.
1. Material culture. 2. Technology. I. Vannini, Phillip.
GN406.M34918   306—dc22   2008025237
ISBN 978-1-4331-0302-5 (hardcover)
ISBN 978-1-4331-0301-8 (paperback)
ISSN 1528-610X

Bibliographic information published by **Die Deutsche Bibliothek**.
Die Deutsche Bibliothek lists this publication in the "Deutsche
Nationalbibliografie"; detailed bibliographic data is available
on the Internet at http://dnb.ddb.de/.

Cover photograph by Laura Lee Gerwing

© 2009 Peter Lang Publishing, Inc., New York
29 Broadway, 18th floor, New York, NY 10006
www.peterlang.com

All rights reserved.
Reprint or reproduction, even partially, in all forms such as microfilm,
xerography, microfiche, microcard, and offset strictly prohibited.

CONTENTS

Preface — vii
Eugene Halton

Introduction — 1
Phillip Vannini

PART 1: WAYS OF KNOWING THE MATERIAL WORLD

1. Material Culture Studies and the Sociology and Anthropology of Technology — 15
Phillip Vannini

2. Actor-Network Theory: Translation as Material Culture — 27
Grant Kien

3. The Social Construction of Technology (SCOT): The Old, the New, and the Nonhuman — 45
Trevor Pinch

4. Material Culture and Narrative: Fusing Myth, Materiality, and Meaning — 59
Ian Woodward

5. Material Culture and Technoculture as Interaction — 73
Phillip Vannini

PART 2: ETHNOGRAPHIC STRATEGIES OF REPRESENTING THE MATERIAL WORLD

6. From Embodied Ethnography to the Anthropology of Material Culture: Gaming in the Field — 89
Mélanie Roustan

7. On Driving a Car and Being a Family: An Autoethnography — 101
Chaim Noy

8. The Screen Deconstructed: Video-Based Studies of the Malleable Screen — 115

Dylan Tutt and Jon Hindmarsh

9. Technologies of Consumption: The Social Semiotics of Turkish Shopping Malls — 131
Tanfer Emin Tunc

10. Cultural Phenomenology and the Material Culture of Mobile Media — 145
Ingrid Richardson and Amanda Third

11. A Grounded Theory Approach to Engaging Technology on the Paintball Field — 157
Ariane Hanemaayer

PART 3: ETHNOGRAPHIC STUDIES

12. What Gardens Mean — 171
Chris Tilley

13. Making It, Not Making It: Creating Music in Everyday Life — 193
Bryce Merrill

14. The Death of the Clinic: Domestic Medical Sensoring — 211
Patrick Laviolette

15. The Zapper and the Zapped: Microwave Ovens and the People Who Use Them — 229
Tina Peterson

List of Contributors — 245

Index — 251

Preface

Eugene Halton

As part of a research project for what would become my book, *The Meaning of Things*, I interviewed a Chicago high-rise apartment dweller whose living room consisted of almost no furniture and over 300 flowering houseplants. There were 200 fuchsias alone, tightly packed in open bins with a watering system, near the wide windows overlooking Chicago's Museum of Science and Industry in Hyde Park. His living room also held various spice plants that he brushed while we walked around the room, releasing a tropical forest of scents.

The high-rise was selected as the location for interviewing a roughly "upper middle-class" sample for the pilot study in 1976, and each floor had a similar identical architectural design. It made it even more interesting that I had been in apartments with the same layout on other floors, and none were remotely similar to this. What might this man's stuff tell us about material technoculture?

When I asked him about his special material possessions, he spoke about his love for the plants, about how they kept him in touch with nature in the middle of the city, his membership in some horticultural societies, and about the status hierarchies of plant keeping. It turned out that "the common coleus," as he put it with a slight look of disgust, was low status, kept by people who seemed to him to be lowbrow, merely following the late 1970s trend of keeping houseplants, whereas his houseplants were high status. It was all fascinating sociology, to see and sense his living symbols of social relations. He was part of the houseplant trend, but had found a way to distinguish himself from it as upper status.

This man worked in philanthropy, rubbing elbows with the rich on a daily basis. He then returned to his high rise haven of high-status flowering plants, a veritable forest of symbols reflecting a disdain for the hoi polloi below him. But he also mentioned as one of his special possessions a Lucite-encased fragment of a cadmium control rod from the world's first nuclear reactor, built nearby on squash courts under the University of Chicago's Stagg Field football stadium bleachers, as part of the Manhattan Project during the Second World War. It produced the world's first self-sustained nuclear chain reaction on December 2, 1942, as Enrico Fermi's "Chicago Pile #1." Here was the first objectification of the atomic age, the first building block of nuclear bombs, preserved as a little living room knickknack, a souvenir.

As I reflected on his possessions after the interview, it struck me that if this apartment were to be preserved a thousand years from now, an archaeologist bumping into it would find no plants, perhaps at most only shelving, watering pipes, and timers for them. Consulting the newspapers of the time, the archaeologist might infer "meth lab" or indoor marijuana growing equipment, some drug business operation, but probably not fuchsias grown for aesthetic and nonutilitarian purposes. Material objects do tell stories, though usually not the whole story. But that archaeologist would still be able to date an artifact from the absolute ground zero of the atomic age, the radioactive rod. What would your possessions, detached from the meanings they hold for you, tell that same archaeologist a thousand years from now?

The atomic age: consider how archaeologists and historians date the swaths of history by the things that humans fashion or use: the Stone Age, the Bronze Age, the Iron Age. And these traditional categories could be followed by the Steel Age, the Atomic Age, and now, perhaps, the Silicon Age. These "Ages" are attempts to picture human prehistory and history through material indicators. Or perhaps it would be better to say through material indicators that last. My respondent's plants would not leave many physical traces and probably no traces of the social relations in which they involved him.

Indeed one of the problems with the "hard-thing" ages of humankind approach to human development is precisely the way it overvalues stuff that lasts and undervalues social practices involving symbolic communication, or soft textile cordage, traps, baskets, shelters, and so on. The soft technologies do not survive, the hard ones do, and so tendencies toward objectivism end up defining a people "Paleolithic" or "Neolithic," based on surviving stone tools. Hence the "hard-thing" view of history and prehistory is itself a mirror of the modern age of materialism and its idealization of the thing.

Archaeologists have only fairly recently begun to correct the tendency to overvalue *homo faber*, the human maker, and undervalue *homo symbolicus*. But this also raises questions concerning the place of materialism and technique in everyday life. Until recently, the ability to use tools was taken as a defining feature of human culture. But Jane Goodall's chimps and a variety of other animals have demonstrated abilities to fashion, use, and transmit tools, suggesting that either culture is broader than humans or that it is abilities other than tool use that mark human culture. Hard tools, though useful indicators, simply do not record the most important steps in the development of symbolic intelligence. The primary technology that transformed us into the category of humanity was the emerging human body, bootstrapping its way through forms of touching, empathic communication, and vocalization, using its flexible brain to attune itself to the inner and outer voices of nature.

Humans emerged in a world that was alive, a biospiritual, signifying world that engaged the ritualizing creature into its emergence. And material technoculture was already there, enabling humans to expand the range of what

Paul Shepard (1998a [1973]) called *the sacred game*, to catch larger game, cook otherwise inedible plants, to travel to remote places and climates, and even to savor, select, and spread flowering plants for their beauty.

The transition from hunter-gatherer life to that of agriculturally based civilization some 12,000 or so years ago was a great watershed of consciousness, not only radically altering the relation to the living environment, but also producing the origins of materialism. Domestication meant that humans began to surround themselves with stuff permanently, instead of foraging for it nomadically. Our world is its legacy.

One of the dubious distinctions of civilization was the introduction of poverty as well as property. The city, as focal point of civilization, elevated an *anthropocentric mind* to the center of human consciousness in its invention of divine kingship and of personified deities, its development of standing military organizations and bureaucratic institutions, its domestication of wild animals and plants, and its rational organization of them and other goods as commodities, capable of being valued for exchange. Cities made a lot of stuff, and control of that stuff became the stuff of power. Lewis Mumford (1967) has described the origins of the city and its institutions as the first *megamachine*, which consisted of mostly human parts. Civilization, as megamachine, is thus at its core implicated in the radical proliferation of material culture and technoculture in everyday living.

But the transition concerned more than *having;* it cut to the quick of *being*, more than I can outline here (see Halton 2007; Mumford 1967; Shepard 1998b [1982]). One sees this transformation even in the term *culture*, which derives from the Latin *colere*, meaning to till, cultivate, dwell or inhabit, and which in turn traces back to the Indo-European root, **Kwel-*, meaning to turn round a place, to wheel, to furrow. The term culture originates in a conception of cultivation, reflecting this changed relation to controlling the plants, animals, and environment. The original meaning also implies the plow, and with it, technology. Manuring, plowing, and transforming land through domestication and agriculture may thus be implicated in **Kwel-*, even as "the culture and manurance of minds"—as Francis Bacon expressed in 1605—showed the original earthy agricultural sense, as well as the emerging transferred use to mind and meaning. In this sense one might say that the culture of this man's apartment and the plants he kept went back to the origins of the word culture itself. Similarly, the term *material*, as in materialism, reveals an unexpected history, being rooted in a term for the base of the trunk of the tree, the life-giving *mater*, or mother. Strange, isn't it, how the term *materialism* came to signify inanimate things instead of the basis of life-giving itself?

A further contraction of consciousness occurred, from *anthropocentric mind* to that of the modern worldview, or what I call *mechanicocentric mind* (Halton 2007). Its chief model was the clock, which, since the fourteenth century, had become a dominant symbol in Western consciousness, reshaping and rationalizing daily life, work discipline, and the very conceptions of time and space (Mumford 1970, 1986; Thompson 1967). The universe itself became redefined as a vast clockwork. As Kepler put it, "My aim is to show that the heavenly machine is not a kind of divine, live being, but a kind of clockwork" (Crosby 1997:84). Who would think that what has long since been an everyday object could have impelled such revolutionary changes into being?

To call a clock "material culture" is to draw attention to how culture manifests itself in communicative practices, which include language, beliefs, and skills, but also how it includes material embodiments as well. A traditional wristwatch, for example, is a microcosm of global culture, encoding a combination of the Babylonian base 60 counting system, the Greek decimal system, Arabic numerals originally developed in India, and two divisions of 12 hours each, deriving from ancient Egypt. A clock is a material object and a technological product; it is also a communicative sign involving the skill one needs to read it, the language of numbers, words, or symbolism needed to decipher it, and the belief not only that it indicates time, but that abstract clock time accurately represents time.

A simple clock is then simultaneously a manifestation of material culture and technoculture. But it can be any number of other things as well, for things may not always be what they seem. Just as the clock came to symbolize the modern mechanical universe, a timepiece can symbolize personal memories, status aspirations (as my students testified to when asked whether they would want to have a Rolex watch); the same applies to a gift from loved ones or co-workers, or even a trophy (as the one that I once won in a track meet, tucked away in the back of my desk). The meanings of things are various, and finding out what they are requires a variety of approaches, from simply asking people what their things mean or observing how they use or do not use them, to backtracking their history, or contextualizing them in broader cultural context.

Material culture and technoculture not only provide openings to study culture, but raise questions about contemporary materialism and technology more generally as well. Consumption is clearly a driving force on the globe today, powering economies, promising identities, providing a cornucopia of commodities. Technoculture is at its center, both in material devices and in the ideas they communicate about how what one has affects what one is and can be. Though Emerson said more than a century and a half ago that "things are in the saddle, and ride mankind," the ride has only galloped ever quicker. Material technoculture is in the saddle, riding with something like the speed of Intel cofounder Gordon E. Moore's law of exponential growth every 2 years. Computers, cell phones, all of them make a world like Alice's (from *Through the Looking Glass*), where "it takes all the running you can do to keep in the same place."

But why should things run us? Isn't technology a means, like any tool, for living? Shouldn't material objects be materials for life? The problem of materialism is not whether to have materials for living, but in allowing them to become goals in themselves.

A device, when correctly used, is a means to human purposes, ultimately a pragmatic means to the good life. Clearly, contemporary devices of technoculture, such as the computer I am typing this on, can serve to enhance our lives. Yet when misused, a device can become a goal unto itself, as when a cell phone or video game dominates a teenager's (or an adult's) life.

What if we consider devices as slaves that ought to be dominated? The Czech word *robota* means slave-like labor. Electronic devices are robotic conveniences, our dominion as masters over them assures that they serve us and not vice versa. To even consider the devices of technoculture as slaves at first sounds so politically incorrect. A slave is a mere *means*, which is why human slavery is evil, in treating fellow humans as means instead of beings with their own ends. But remember the old term servo-mechanism? We did not eradicate slavery in modern life, but only transferred it onto devices, which seem ever increasingly to dominate everyday life. Hegel's dilemma of the master and slave relationship is still there, but transformed. The danger of relying too heavily on automatic culture is that we become dominated by it, and the original goal of technology as serving human autonomy becomes reversed: we become more automatic as the devices seem to become more "autonomous."

When the means of life found in material objects, such as the devices of technoculture, become ends that usurp the good life, the result is a dehumanized end that could be called *the bad life* (or if you do not like that, *the not so good life*). Mumford (1963) identifies this tendency to unchecked expansion of technoculture with the megamachine in its ancient and modern variants, what he termed *authoritarian technics*. In contrast, objects and techniques kept in their limited place as means to the good life are what he termed *democratic technics*.

I view such limited use of material culture as *instrumental materialism*, the use of objects as means to realize goals, in contrast to *terminal materialism*, or the treating of things as ends instead of instruments of life (Halton 2008:227).

Hence mastery of technical devices, the treating of them as servants and means, involves using them as instruments for self-determination, autonomy, and the common good. It also involves limiting their use, knowing when, and even how, to shut them down (as the story of *The Sorcerer's Apprentice* illustrates). Such mastery of the thing allows one to be vulnerable to life, rather than be armored off from it by the things one surrounds oneself with.

So consider that Chicago apartment with the fuchsias and nuclear control rod. It was just a living room, one doubtlessly long since moved from or disassembled. But in this little living space was a micromaterial history of technoculture. Here was the fruit of citifying domestication, embodied in the array of high-status flowering plants, none of which were there to be eaten, only to be enjoyed and admired.

Here too was the clock, symbol of the modern rational-mechanical universe, embodied in the lighting and watering timer devices of this floral-mechanical living room, signifying liberation from diurnal cycle and season. So too was the control rod both a kind of radioactive clock and a sign of human liberation from solar energy through nuclear fission. We extracted the rational-mechanical elements of ourselves, projected it onto the heavens, measured it with precision, and declared the physical universe a vast clockwork and measure of all things. We fell into materialism, and more recently into technomania, elevating the automatic aspects of life while too frequently losing sight, it seems to me, of our place in the communicative community of life.

I have given you a "big picture" way of looking at this man's things, which allowed me to show a history of material culture and technology through them. I could also have examined other dimensions, how, for example, his early childhood in a wealthy North Shore Chicago suburb and his mother's garden might figure psychologically into his plants and their meanings. But this is simply to say that there are any numbers of other ways through which to view the meanings of things and technoculture, as you will see amply demonstrated in this book.

References

Crosby, Alfred. 1997. *The Measure of Reality: Quantification and Western Society, 1250-1600.* Cambridge: Cambridge University Press.

Halton, Eugene. 2007. "Eden Inverted: On the Wild Self and the Contraction of Consciousness." *The Trumpeter*, 23:45-77. Available at:
http://trumpeter.athabascau.ca/index.php/trumpet/article/view/995/1387

———. 2008. *The Great Brain Suck: And Other American Epiphanies.* Chicago: University of Chicago Press.

Mumford, Lewis. 1963. "Authoritarian and Democratic Technics." Lecture, The Center for the Study of Democratic Institutions, New York, January 21-22. Available at:

http://www.primitivism.com/mumford.htm

———. 1967. *The Myth of the Machine, Volume 1: Technics and Human Development.* New York: Harcourt, Brace, Jovanovich.

———. 1970. *The Myth of the Machine, Volume 2: The Pentagon of Power.* New York: Harcourt, Brace, Jovanovich.

———. 1986 [1934]. "The Monastery and the Clock." Pp. 324–332 in *The Lewis Mumford Reader*, edited by in Donald Miller. New York: Pantheon.

Shepard, Paul. 1998a [1973]. *The Tender Carnivore and the Sacred Game.* Athens: University of Georgia Press.

———. 1998b [1982]. *Nature and Madness.* Athens: University of Georgia Press.

Thompson, E.P. 1967. "Time, Work-Discipline, and Industrial Capitalism." *Past and Present*, 38:56–97.

Introduction

Phillip Vannini

The drive to work was awfully cumbersome today. Two accidents—each with its own mile-long backup—treacherous road conditions, and heavier-than-normal Monday morning traffic led to a very late arrival to the office and in a lot of lost patience. The primary cause behind all this? A wildly unusual mid-April snowfall that made roads slippery and drivers edgy.

Now, why in the world—you must be wondering—is this worthy of concern in the opening of this book? Well, because—as it turns out—this is precisely the subject matter of this volume. No, not the snow itself, but rather the technological and material character—or in other words, materiality—of everyday life of which the snow, the roads, the size and weight of the vehicles driving on them, the technical skills of drivers, the quality of tires, the density of traffic, the meanings of driving, the road infrastructure offering (or not) commuters alternative routes—and the availability of maps and GPS systems to find out about such routes—as well as the air temperature (only to mention a few) are great examples. Indeed, this is a book about what makes everyday life possible (and at times, like this morning, difficult) stuff. Or, in more technical terms, material culture and technology.

So what kind of stuff are we talking about when we say "material culture" and "technology"? In a way, it really depends on whom you ask. If you asked lexicographers they might suggest that "technology" refers to the "practical application of knowledge" or a "capability given by the practical application of knowledge" and even a "manner of accomplishing a task" (*Merriam-Webster's* online dictionary). According to the same folks "material," on the other hand, refers to something "relating to, derived from, or consisting of matter" (we won't ask them about "culture" for now). In contrast, anyone "off the street" might give you a simpler set of definitions. Technology for them might refer to machinery, gadgetry, or devices with which one accomplishes instrumental tasks, whereas material could simply denote objects or things. This is simple enough so far: both the erudite writers of our dictionary and Joe and Jane Average would agree that technology is about doing, knowing, and using objects and that materiality is about the character of those objects. Given that the concept of culture, broadly speaking, refers to practices, bodies of knowledge, ways of engaging the world, and so forth, we might be tempted at this point to claim that to speak of technology or to speak of material culture is basically the same thing. But we should not. At least not just yet, and at least not in those terms; as the good academics that we are, both you and I, it behooves us to complicate things a little bit before we reach a conclusion. So shall we?

Material Culture and Technoculture

To suggest that the subject of this book is both technology and material culture is to imply that these two topics at the very least have something in common, and that perhaps they are even somewhat of the same entity (cf. Eglash 2006). Indeed it would be wrong to disagree with the validity of that implication. To realize this better it is best to situate our comprehension in the pertinent academic research and theoretical traditions rather than in the common or the lexical understanding of technology and material culture. In doing so we will have a deeper grasp of what this book is about. Now, to get there one can find at least three such traditions to borrow from traditions clearly tangible if you were here with me in my office and could see the three neatly stacked piles of books and articles crowding my desk.

The first stack, or tradition, is about the contemporary study of material culture, or as some refer to it, modern material culture studies (e.g., Buchli 2002b; Miller and Tilley 1996; Tilley et al. 2006). Modern material culture studies attempt to rediscover the significance of objects not only in terms of their role in economic exchange, but also (and more importantly) in terms of their cultural role (see Appadurai 1986; Douglas and Isherwood 1979; Gell 1998; Strathern 1988)—a role historically considered secondary in most social scientific disciplines traditionally more interested in what is "behind" (e.g., values, beliefs, mind, collective consciousness, social structures, etc.) material objects (Buchli 2002a; Knappett 2005; Miller 1998). As Miller and Tilley (1996) have stated, material culture studies is an interdisciplinary field—though it is obviously deeply rooted in social anthropology and ethnoarchaeology.[1] Regardless of this tendency the contemporary study of material culture is an open discipline, both theoretically and methodologically, with a common concern: processes of objectification, through which humans shape, and are shaped by, the materiality of life (see Miller 2005; Tilley 2001, 2006; Woodward 2007). More on this later.

The second stack of books and articles on my desk is about the social aspects of technology. This is a very diverse pile that comprises writings on the anthropology of technology (e.g., Ingold 2000; Lemonnier 1993; Pfaffenberger 1988, 1992), science and technology studies (STS) (e.g., Latour 2007), the philosophy of technology (Scharff and Dusek 2003) and cultural theory on technology (e.g., Haraway 2003; Penley and Ross 1991), communication and cultural studies (Carey 1989; Cowan 1983; Du Gay et al. 1997; Fischer 1994; Silverstone and Hirsch 1992; Slack and Wise 2007), the sociology of technology and science (e.g., Clarke and Olesen 1998; Star 1995), the social construction of technology (SCOT) (e.g., Bijker, Hughes, and Pinch 1989;

Bijker and Law 1992; MacKenzie and Wajcman 1999) and within it especially the subfield of technology users (Oudshoorn and Pinch 2003).[2] Despite their diversity most of these scholars would agree on the idea that social relations in all societies are heavily mediated by technological arrangements, and that therefore technology (as a form of social organization) is a key player in society and culture.

Despite the similarity—or at least the contiguity—of the concerns of material culture studies and technology studies, not much cross-proliferation has seemingly taken place (for recent exceptions see Eglash 2006 and some of the literature cited therein; Pinch and Swedberg 2008). One might argue that in part this is due to the different geographical origins of these fields—the first being decidedly more British and in lesser part French, and the latter being decidedly more American. Or one might argue that it is instead due to the remnants of disciplinary boundaries (with material culture studies being decidedly more anthropological and technology studies being definitely more sociological). But whatever the causes may be, we can safely agree that such boundary is the result of accidental practice instead of motivation and planning (see chapter one). Material culture studies and technology studies have much in common, and for this reason they should be drawn in closer dialogue. This book is written with that intention.

As each of the chapters in the following pages shows, to study material culture is to study the technological underpinnings of culture, and to study technology is to study the material character of everyday life and its processes of objectification. What is central to such a view is an understanding of sociality and culture as a form of *making, doing,* and *acting* and an understanding of the world as a material presence apprehended by humans through pragmatic, sensuous intentionality. In comprehending culture as deeply shaped by techne—that is, craft, skill, creativity—and in viewing social interaction as a process rich with material properties we do not intend to either reintroduce antiquated notions of instrumentalism or essentialism. Rather, we simply intend to remark on the importance of treating everyday life as an active form of negotiation—a form of work as it were—that engages the colors, the textures, the tastes, the fragrances, the sounds, the temperature, the kinaesthetic movement, and the practical and symbolic value of the stuff that makes up life.

If bringing together the tradition of material culture studies and technology studies is a key concern of this book, so is achieving that goal through methodological and epistemological means that expose the meaningfulness and polysemy of materiality, and the potential of technological relations for shaping culture (and being shaped by it). For us what that means is ethnography: the subject of the next reflection. But before we get there it is important to realize that we have come full circle in our own understanding of technology and material culture, and in claiming that the semantic differences between these expressions are more the result of putative scholarly practice than ontological reasons. Thus, throughout this book, I and every other author will refer to *material culture* and *technological culture* (or *technoculture*) interchangeably. These expressions point to an emergent process consisting of the interaction between human actors and nonhuman actors—all acting with their strategies and techniques, endowed with material properties. Also, by using interchangeably the words "material objects," "things," "technics," "technological devices," or similar expressions we will refer to the same thing: the resources (cf. Gibson 1979; see also Van Leeuwen 2005) that actors use for instrumental and symbolic purposes. In fact, we will view the difference between instrumental and symbolic purposes (and the related dissimilarity between function and style) to be hindering more than helping our agenda, and for this reason we will explicitly blur the boundaries between action and communication. When different concerns and arguments force some contributors to favor the use of certain expressions over others, we ought to keep in mind that their lexical choices are motivated by their need to treat different empirical subject matter with attention to detail, rather than to reify categories by erecting unnecessary boundaries among them. With that in mind, each chapter of this book will feature various approaches and highlight different angles of our common subject matter. Indeed variety and diversity are the strength of any edited book. Yet the chapters that follow have their origin in the shared understanding that technology is never in the things themselves, in materiality alone, in the techniques and strategies of makers or users alone, in the selves and collective identities of makers or users alone, in the discourses encompassing the interaction between human and nonhuman actors alone, but rather in the process whereby all those entities interact and give form and content to our world. To speak of technology, therefore, will entail speaking of technoculture or material culture. And to speak of materiality, therefore, will entail speaking of material culture or technoculture.

The third and last stack sitting on my desk consists of books on ethnography. Traditionally rooted within both cultural anthropology and classical urban sociology, ethnography is now one of the most common research strategies across the social sciences, and one that is currently enjoying an impressive outburst in creativity, scope of applications, and diversity. While there are many types of ethnography, its defining characteristic resides in the researcher's attempt to understand realities from the perspective of those he/she wishes to study. In practice, this form of epistemological relativism translates into data collection methods such as (different combinations of) participant and nonparticipant observation, conversations and interviewing, analysis of records, texts, material objects, and reflection and introspection. In its focus on mundane practices of social actors, attention to context, and emphasis on agency and interaction (Adler, Adler, and Fontana 1987), ethnography is the everyday life research strategy par excellence. While there are other research strategies that could direct us on the everyday life aspects of material culture and technoculture, in this book we focus specifically on ethnography alone because we find that its application to the subject matter of our field—while already prolific and successful—requires reflection and further development.

In particular, we aim for methodological reflection that can allow us to surpass—or at least be more cognizant of—the limitations of traditional ethnographic research strategies in relation to material culture and technoculture. An example ought to shed light on the nature of these limitations. Suppose—to return to the opening of this introduction—that we wished to study the meaningfulness of an unusual mid-spring snowfall in relation to the value of mobility and the technological infrastructure of roads in a particular geographical area. What information could an ethnographic research design provide us with? Observation from the roadside or the cocoon of your car—if you are lucky enough to be caught in traffic as it is happening—might yield impressions, reflections, and experiences of driving in such conditions. A later search through publicly available data on traffic and road infrastructure, as well as on historical weather records, might give us further knowledge to put our observations in proper context. But those data—even when rich in volume and detail—could be insufficient for our scope. As most ethnographers do, we might then decide at the end of the day to interview drivers who were caught in the snowstorm. And here is where both our methodological potential and problems might begin to be obvious. Even assuming that a sample of drivers is promptly available and enthusiastic enough to dedicate sufficient interview time to us (and this may be difficult, given how reticent some people may be to invest time on reflecting on such mundane matters), we will inevitably run into the difficulties of gathering interview data that are sufficiently insightful, or in other words not thick enough, for our purpose. This is no one's fault; after all who—even among the most eloquent and articulate of us—would have detailed answers for interview questions directed at uncovering the meanings of unseasonal (or seasonal for that matter) snow precipitation, the values underlying highway mobility, or the significance of studded winter tires. Even in the best case scenario ethnographic interviews of that type might yield either the kind of practical information that applied researchers and policymakers can best obtain otherwise through large surveys (the kind of information that cultural scientists find uninteresting) or perhaps superficial insights that are useless and without the kind of in-depth theoretical interpretation, which inevitably ends up losing sight of the raw interview data themselves—and thus makes their ethnographic collection pointless.

Undoubtedly, limitations of this kind are not insurmountable. While it may be extremely difficult to make mute material objects like spring snow "speak" (see Hodder 2003; Tilley 2001), it would perhaps be a best strategy altogether to partly underemphasize symbolism and logocentrism and focus instead on other, less understood modes of objectification (Tilley 2001), communication (Knappett 2005), and representation (Vannini, Hodson, and Vannini 2008). Thus, deemphasizing traditional ethnographic concerns with the values "hiding" behind actions, the mind underlying embodied interaction with objects, the discourses surrounding action, the teleological purposes buttressing technology, and the symbolism overriding iconic and indexical meaning might allow us to recenter ethnographic methodology in a way that is more consonant with the subject matter of material and technoculture research. Such is a quest for a methodology that does not privilege communication at the expense of action, being at the expense of doing, consciousness at the expense of materiality, and speaking at the expense of making. Such new ethnography places techne as much as ethnos at the core of its concern. Through multiple approaches, through different foci, and with various scopes, the contributors to this book reflect on the present and future potential of ethnographic methodology—itself a technology affording opportunities and constraints—to further develop our knowledge on material culture and technology.

An Everyday Life Approach to Everyday Life

All studies of everyday life take the mundane as its subject matter, but often neglect to take a methodological approach that is also grounded in the practices of everyday life. To explain this let us take the example of speed bumps. Speed bumps—or sleeping policemen according to Latour (1992)—objectify normative values and systems of authority by causing drivers to slow down as they approach them. Because police cannot be everywhere at the same time, they and architects delegate the task of slowing down to speed bumps. In carrying out their scripts speed bumps faithfully manifest their agency, Latour explains. Simple enough, right? Any keen student or scholar of everyday life would promptly recognize a chapter on this subject as worthy of cataloguing under their favorite category next to microwave ovens, waffle-makers, planted flowers, and video games. But in the end what makes a speed bump more or less mundane "everyday life-like" than geothermal energy, or maybe something like nutrification? The fact that, perhaps, we encounter speed bumps more or less every time we head to the mall whereas we have hardly even hear of geothermal energy or nutrification? This explanation cannot be sufficient.

The idea that a social scientific study of everyday life takes into account the things that "we" encounter more or less daily is partial at best and wrong at worst. For starters, the expression "everyday life" seems to imply that only the things that actually happen *every* day should count as its subject matter. However, note that it takes into account not only every day but also *everyone's* life! When is the last time *you* used a waffle-maker? Or played a video game? The expression "everyday life" may be understood as standing for somebody's daily realities that we are discussing, but it's not somebody else's. If you live in the countryside, for example, you may never encounter speed bumps until you go to town. Does that make your everyday life less normal—and you, as well, arguably—than your urban counterpart? Obviously a definition of everyday life studies based on the temporal frequency of mundane behaviors, or the convenient exclusion of some people from some behaviors, cannot be but unsatisfactory.

Second, should we not be concerned with our own—as researchers—mundane practices beside those of our informants? In other words, shouldn't our (again, as students and scholars) everyday life count? Roustan (chapter six), for example, had hardly ever played video games until she embarked on her fieldwork. Can she thus logically call her investigation one of everyday life? If so, what is the true difference, if any, between the practice of playing video games and offering sacrificial gifts to the mountain gods? Both, after all, are entirely foreign to the researcher, and both have deep consequences for everyday living—whether that is missed opportunities to hang out with one's girlfriend or a healthy daily relationship with omnipotent mountain gods. What really matters, we start to realize, is not really the topic, but perhaps something else.

What makes (or breaks) the study of everyday life—we argue in this book—is the very same "attitude" that we, as human beings, have in everyday life. What that attitude might entail is entirely dependent on whom you ask. For Woodward (chapter four) everyday life has the characteristics of story. For Pinch it has the feeling of an evolving (chapter three) historical outcome of "lessons" learned from the past. For Richardson and Third (chapter ten) an everyday "attitude" has the feel of sensual experiences. For Tutt and

Hindmarsh (chapter eight) instead it is based on a rather "taken for granted" approach to doing things, an approach that is based on people's routine methods for getting things done. For me (chapter five) the attitude of everyday life is performative and dramatic, that is, based on action and interaction. And so on. While there is no ultimate answer on what *the* characteristic attitude of everyday life is—and that is why there are several theoretical and methodological perspectives available—all the contributors to this book agree that an approach to everyday life has come to terms with the stuff that makes up daily living, and it has to do so in the typical ways in which we encounter this stuff.

Thus, if we do so, we soon come to realize that without taking into account context, the agents (human and nonhuman) involved, their purposes, and all the various contingencies surrounding their interaction we understand very little about everyday life and risk coming to very partial and at times downright erroneous solutions. Such, in the end, should be the spirit of the study of everyday life. And this is also the spirit, we find, of ethnography. As Roustan explains in her chapter, it is the ethnographic spirit that allows her to be skeptical of the claim that gaming has "virtual" or somehow less-than-real characteristics. As Noy narrates, it is contingency and relationships that shed light on why he decided to speed over a sleeping policeman. And as Peterson finds, a microwave oven is not a product that necessarily revolutionizes cooking, concept of gender, and family relationships, but instead, a technic that at least at times and for some people, provides users with occasion to fight wars with makeshift marshmallow action figures. What truly distinguishes the study of everyday life, in the end, is not that it is mundane stuff, but rather that it is as naive, as curious, as inquisitive, as involved, as unpredictable, as odd, and as crafty as everyday life itself. And guess what? A name for this attitude, and the practices it manifests itself in, might as well be what Greek philosophers called "techne."

Outline of the Book

Following this brief introduction, Grant Kien, Trevor Pinch, Ian Woodward, and I review the basic theoretical assumptions and key conceptual tools that the theoretical perspectives of the anthropology of technology, modern material culture studies, the sociology of technology (chapter one), Actor-Network Theory (chapter two), the social construction of technology or SCOT (chapter three), narrative theories (chapter four), and interactionist and performative perspectives (chapter five) bring to bear upon the study of technology and material culture. The five chapters making up part one of the book work in part as analytical reviews of the current state of each perspective—as it pertains to the study of material culture as technology—and in part as brief literature reviews of notable theoretical works. Each of these chapters lays out the basic framework for understanding how the material/technological world of everyday life can be known ethnographically.

Part two is less theoretical, and more explicitly methodological and empirical. Each chapter in this part outlines a different ethnographic tradition in both abstract terms and concrete application through a case study.

The first chapter in part two introduces the very characteristics of an everyday life approach and its potential for demystifying very un-everyday-life-like attitudes. Building on the tradition of embodied ethnography Roustan examines how video game players engage their leisure tools through bodily skills, habits, and techniques that demand high degrees of familiarity and daily practice. Subsequently, in a chapter that is just as grounded in the daily realities of dialogue and routine, Noy, borrowing from the young but booming autoethnographic tradition in material culture studies, reflects on interaction inside, and with, the family car. Much can be missed in recollection and description, as we learn in chapter eight. Drawing upon the tradition of ethnomethodology, conversation analysis, and visual ethnography, Tutt and Hindmarsh reflect on how users make sense of, and engage, types of monitor screens in a variety of work activities. The subsequent chapter, chapter nine, concentrates on social semiotics. Through her analysis of ethnographic data collected at, and about, American-style Turkish shopping malls, Tunc demonstrates how mass-marketed consumer items work as semiotic resources in both an expressive (symbolic) and instrumental (technical) way. Exploring the research strategy of cultural phenomenology while borrowing extensively from visual ethnography Richardson and Third, in chapter ten, focus their empirical attention on Australian teenagers' use of cell phones, PDAs, mp3 players, and handheld game consoles and the significance of these objects as perceived, sensually, by their users. Finally, the last chapter in part two discusses the use of one of the most common qualitative strategies across the social sciences: grounded theory. The empirical subject matter of Hanemaayer's chapter is the culture of paintball fighters and their relation with their weaponry.

Part three of this book features longer empirical studies working as illustrations of the various ways—theoretical and methodological—in which ethnographers can study material culture and technology. Each chapter in this part liberally borrows from a variety of traditions, blending different—but compatible—elements. Part three begins with Tilley's theoretically eclectic ethnographic study of the significance of urban and suburban gardens and the practice of gardening in England. In chapter thirteen Merrill combines interactionism with SCOT in his analysis of the practice of home-based musical recording. In chapter fourteen Laviolette examines the colonization of the mundane and the body by the hand of the introduction of health care technologies in domestic settings. Finally, in chapter fifteen Peterson applies symbolic interactionism to her interpretation of microwave usage. As these studies show, ethnographic research that carefully engages the materiality of social interaction—without abandoning itself to radical forms of constructionism or essentialism (cf. Ingold 1996)—can demonstrate how material culture, or technoculture, is ultimately none other than a practical, everyday way of doing things with things.

Notes

1. And even though it is an international field one might even note—at least by judging from the *Journal of Material Culture*, the flagship journal of contemporary material culture scholars—that the core of this field can be found in the UK, and even more precisely in such schools of thought as the University College of London (see Buchli 2002a) and Cambridge University (see Henare, Holbraad, and Wastell 2007). Other geographical concentrations can, however, be found in France and in particular within the *Matière à Penser* School, and in the United States around the *Winterthur Portfolio* journal. It is the British and French tradition, however, that primarily influence my treatment of material culture in this introduction.
2. Even though this is a diverse bunch of scholarship, one cannot help but noticing in it a more distinctly American flavor—and in part also Northern European—which is especially obvious within STS, SCOT, and the field of the history of technology that has coalesced around the journal *Technology and Culture*.

References

Adler, Patricia, Peter Adler, and Andrea Fontana. 1987. "Everyday Life Sociology." *Annual Review of Sociology*, 13:217-235.
Appadurai, Arjun (Ed.). 1986. *The Social Life of Things*. Cambridge: Cambridge University Press.
Bijker, Wjebe, Thomas Hughes, and Trevor Pinch (Eds.). 1989. *The Social Construction of Technological Systems: New Directions in the Sociology and History of Technology*. Boston, MA: MIT Press.
Bijker, Wjiebe E. and John Law (Eds.). 1992. *Shaping Technology/Building Society*. Cambridge, MA: MIT Press.
Buchli, Viktor. 2002a. "Introduction." Pp. 1-22 in *The Material Culture Reader*, edited by Viktor Buchli. New York: Berg.
—— (Ed.). 2002b. *The Material Culture Reader*. New York: Berg.
Carey, James W. 1989. *Communication as Culture*. New York: Routledge.
Clarke, Adele and Virginia Olesen (Eds.). 1998. *Revisioning Women, Health, and Healing: Feminist, Cultural, and Technoscience Perspectives*. New York: Routledge.

Cowan, Ruth Schwartz. 1983. *More Work for Mother: The Ironies of Household Technology from the Open Hearth to the Microwave.* New York: Basic Books.
Douglas, Mary and Baron Isherwood. 1979. *The World of Goods.* London: Allen Lane.
Du Gay, Paul, Stuart Hall, L. Janes, H. Mackay, and K. Negus. 1997. *Doing Cultural Studies: The Story of the Sony Walkman.* London: Sage.
Eglash, Ron. 2006. "Technology as Material Culture." Pp. 329-340 in *Handbook of Material Culture*, edited by Chris Tilley et al. London: Sage.
Fischer, Claude. 1994. *America Calling: A Social History of the Telephone to 1940.* Berkeley: University of California Press.
Gell, Alfred. 1998. *Art and Agency.* Oxford: Oxford University Press.
Gibson, James. 1979. *The Ecological Approach in Visual Perception.* Hillsdale, NJ: Lawrence Erlbaum.
Haraway, Donna. 2003. *The Donna Haraway Reader.* New York: Routledge.
Henare, Amiria, Martin Holbraad, and Sari Wastell. 2007. *Thinking through Things: Theorising Artefacts Ethnographically.* New York: Routledge.
Hodder, Ian. 2003. "The Interpretation of Documents and Material Culture." Pp. 155-175 in *Collecting and Interpreting Qualitative Materials*, edited by Norman K. Denzin and Yvonna Lincoln. Thousand Oaks, CA: Sage.
Ingold, Timothy. 1996. "Situating Action VI: A Comment on the Distinction between the Material and the Social." *Ecological Psychology*, 8:183-187.
———. 2000. *Perception of the Environment: Essays in Livelihood, Dwelling, and Skill.* New York: Routledge.
Knappett, Carl. 2005. *Thinking through Material Culture: An Interdisciplinary Perspective.* Philadelphia: University of Pennsylvania Press.
Latour, Bruno. 1992. "Where Are the Missing Masses? The Sociology of a Few Mundane Artifacts." Pp. 225-258 in *Shaping Technology/Building Society*, edited by Wiebe Bijker and John Law. Cambridge, MA: MIT Press.
———. 2007. *Re-assembling the Social: An Introduction to Actor-Network Theory.* Boston: New York University Press.
Lemonnier, Pierre (Ed.). 1993. *Technological Choices: Transformation in Material Culture since the Neolithic.* London: Routledge.
MacKenzie, Donald and Judy Wajcman (Eds.). 1999. *The Social Shaping of Technology.* Boston, MA: McGraw-Hill.
Miller, Daniel. 1998. "Why Some Things Matter." Pp. 3-23 in *Material Culture: Why Some Things Matter*, edited by Daniel Miller. London: UCL Press.
——— (Ed.). 2005. *Materiality.* Durham, NC: Duke University Press.
Miller, Daniel and Chris Tilley. 1996. "Editorial." *Journal of Material Culture*, 1:5-14.
Oudshoorn, Nelly and Trevor Pinch (Eds.). 2003. *How Users Matter: The Co-construction of Users and Technology.* Boston, MA: MIT Press.
Penley Constance, and Andrew Ross (Eds.). 1991. *Technoculture.* Minneapolis: University of Minnesota Press.
Pfaffenberger, Bryan. 1988. "Fetishised Objects and Humanised Nature: Towards an Anthropology of Technology." *Man*, 23:236-252.
———. 1992. "Social Anthropology of Technology." *Annual Review of Anthropology*, 21:491-516.
Pinch, Trevor and Richard Swedberg. 2008. *Living in a Material World: Economic Sociology Meets Science and Technology Studies.* Boston, MA: MIT Press.
Scharff, Robert and Val Dusek. 2003. *Philosophy of Technology: The Technological*

Condition—An Anthology. New York: Blackwell.
Silverstone, R. and E. Hirsch (Eds.). 1992. *Consuming Technologies: Media and Information in Domestic Spaces.* London: Routledge.
Slack, Jennifer Daryl and J. Macgregor Wise. 2007. *Culture + Technology: A Primer.* New York: Peter Lang.
Star, Susan Leigh (Ed.). 1995. *Ecologies of Knowledge: Work and Politics in Science and Technology.* Albany, NY: SUNY Press.
Strathern, Marylin. 1988. *The Gender of the Gift.* Berkeley: University of California Press.
Tilley, Chris. 2001. "Ethnography and Material Culture." Pp. 258-272 in *The Handbook of Ethnography*, edited by Paul Atkinson, Amanda Coffey, Sara Delamont, John Lofland, and Lyn Lofland. London: Sage.
——. 2006. "Objectification." Pp. 60-73 in *Handbook of Material Culture*, edited by Chris Tilley. London: Sage.
Tilley, Chris, Webb Keane, Susanne Küchler, Mike Rowlands, and Patricia Spyer (Eds.). 2006. *Handbook of Material Culture.* London: Sage.
Van Leeuwen, Theo. 2005. *Introducing Social Semiotics.* New York: Routledge.
Vannini, Phillip, Jaigris Hodson, and April Vannini. In press "Toward a Technography of Everyday Life." *Cultural Studies* ←→ *Critical Methodologies.*
Woodward, Ian. 2007. *Understanding Material Culture.* London: Sage.

PART 1
WAYS OF KNOWING THE MATERIAL WORLD

1
Material Culture Studies and the Sociology and Anthropology of Technology

Phillip Vannini

Despite the seemingly endless availability of studies on the material objects of everyday life, relatively little research on material culture or technoculture either profitably draws upon the pertinent bodies of empirical and theoretical literature or contributes to the development of the field's key concepts. Without wishing to create or police the boundaries of what is meant to be an interdisciplinary field (cf. Miller and Tilley 1996), in this chapter I attempt to offer readers tools to better situate the study of material culture and technoculture within a broad but focused tradition that spans the fields of sociology, anthropology, and communication and cultural studies. I do so by introducing a few concepts and by subsuming them under the idea of objectification. The idea of objectification, I argue, is at the core of the *materialist perspective* that is advanced in this chapter and volume as a whole.

Though there may be divergent understandings of what a materialist perspective or paradigm entails, it would be safe to state that a materialist perspective does not reduce materiality to an epiphenomenon of social processes, nor does it view materiality as merely another form of language or discourse (Tilley 2006a, 2006b). To explain this idea, let us begin with an example. Suppose you wished to study the social phenomenon of wine consumption. Framed as a study in conspicuous consumption, for instance, or in the social display of taste and distinction, this would be rather easily done. But suppose that your interest was in the actual materiality of wine—rather than its incidental symbolic role in the representation of status differences—as well as in wine-making and wine-consuming as technoculture. Suddenly, as you might imagine, things would become quite more complicated. To begin with, you would need a wide array of resources to make sense of the material properties of a liquid such as wine from the perspective of its taste and related sensations. Subsequently, you would need to couch both wine-making and wine-tasting in the vocabulary of action, experience, and use. Following that, you might wish to expand your scope a bit and reflect on how wine-making and wine-drinking objectify social relations, values, and ideas about food and drink. Such an investigation might lead you to consider the social scripts surrounding the making and consuming of wine, as well as the social semiotic resources people use to identify the act of drinking wine. And finally, you might wish to engage in an even broader reflection on the politics of wine production and consumption, ranging from the ideologies and regimes that regulate those activities to how the process of engaging in both reproduces or perhaps contests and negotiates hegemonies of taste. Quite an undertaking, undoubtedly! Yet, I would assume, you might promptly recognize that such an undertaking would be absolutely necessary for a thorough understanding of the materiality of the social phenomenon of your interest. In this chapter I lay out the broad interdisciplinary foundations that would inform a comprehensive study of material culture and technology such as the hypothetical, above-mentioned research project. Drawing upon the disciplines and fields of material culture studies, the sociology of technology, the anthropology of technology, and in lesser part communication and cultural studies, I survey basic ideas and common concepts useful for the ethnographic study of material culture and technoculture.

The scope of this chapter is threefold. First, I wish to highlight how the different bodies of literature I reference—and other authors of various chapters that comprise this book do as well—often share related ideas and concepts, regardless of their theoretical or disciplinary origin. Second, because such bodies of literature often "belong" to different disciplines or traditions that do not often "talk" to one another, in surveying different bodies of literature and related shared concepts I hope to articulate different traditions and bring them on a common platform. Third, by outlining common ideas and concepts shared across all the chapters of this book, I hope to lay down the theoretical foundations of the entire edifice of this book in one convenient location. Given the limited space allotted here, the purpose of the chapter is not that of providing the reader with an extensive review but instead that of giving the reader a summary and directory for research across different but related fields.

Experience, Choice, Technique, and Use

A key component of the study of material culture in mundane settings is the contextualization of objects in people's lives. Drawing upon a broadly defined phenomenological perspective, the student of material culture ought to take into serious consideration, or even squarely focus one's research design, on the lived experience of materiality. Exemplary of this approach are ecological approaches to the perception of the environment (see Ingold 2000 and 2007) and landscape (Tilley 1997, 2006a, 2006b). These studies and perspectives clearly show how experience is by necessity active and constructive, rather than passive and determined by imprint-like characteristics of external stimuli. The phenomenological focus on individuals and their experiences and practices translates into a variety of study angles and concepts such as choice, technique, and use.

Choice becomes an issue of great import when we, as ethnographers, enter a research site and ask ourselves the classic sensitizing question "how could things be otherwise?" Glance at your computer keyboard, for example, and ask yourself why Q-W-E-R-T-Y are the first six letters on the upper row. Could other letters have been chosen? Are keyboards different anywhere in the world? Have they been different in the past? Why? As Lemonnier (2002:6, original emphasis) observes,

> technical actions under construction as well as changes in technology are in part determined or encompassed by social representations or phenomena that go far beyond mere action on matter, societies seize, adopt or develop certain technical features (principles of action, artefacts, gestures) and dismiss others. It is as though societies *chose* from a whole range of possible technological avenues that their environment, their own traditions and contacts with foreigners lay open to their means of action on the material world.

Technological choices show the relative malleability of the material world and clearly anchor the constantly evolving design and production of technics as well as the evolution of techniques in social contexts (see Dobres 2001; Ingold 1996; Pinch this volume).

A common concept in technology studies that highlights the growing mutual adaptation of individuals to the design, function, and significance of objects—as well as the shaping of the latter in accordance to the needs and values of the former—is that of *domestication* (Lie and Sorensen 1996; Silverstone and Hirsch 1992). Domestication refers to the incorporation of objects into the realm of the mundane, or in other words as Silverstone and Hirsch (1992) have famously put it, to the taming of the technologically wild and the cultivation of that which has been tamed. Studies of domestication show how practices and experiences of objects and techniques become familiar by following an irregular historical path that may lead to adoption and domestication or conversely to abandonment. Whether it is the process of making the right tools for the task at hand (Casper and Clarke 1998), of making sense of the meanings of objects and the process of work (Heath, Knoblauch, and Luff 2000), of cultivating objects as extensions of the self (Csikszentmihalyi and Rochberg-Halton 1981), or even of causing the rearrangement of entire technological systems due to the emergence of unexpected patterns of use (Fischer 1994; Martin 1991), studies of domestication and related processes show how technics are not only subjects driving change and transformation but also objects of such processes themselves (e.g., Moser 2000; Sillar 1996; see also Pinch this volume).

Domestication processes generally show four phases: appropriation, objectification, incorporation, and conversion (Silverstone and Hirsch 1992). In appropriation a technical object or service is sold from producers and distributors to individuals. In objectification (keep in mind that this concept differs from the use of objectification to be more fully discussed later) the principles of the "household's sense of itself and its place in the world" (22) are revealed and displayed. In incorporation objects are integrated into daily routines. And finally, in conversion objects are put into practice and thus begin to shape relationships between users and others. For example, objects may be used to make status claims, symbolize identity, and so forth. As Lie and Sorensen (1996) have remarked, domestication entails different kinds of interaction, such as symbolic work, practical work, and cognitive work. Symbolic work is action directed at shaping and negotiating meanings of objects. Practical work is the development of habits, patterns of use, and familiar techniques. And cognitive work refers to learning and narrativizing (see Woodward this volume).

Another important component of studies of technology use and users is research focused on how users are defined, by whom, and through what means. Regardless of one's theoretical perspective the process of coconstruction of technics, techniques, and technicians is a necessary object of consideration if one wishes to transcend a simplistic, deterministic approach to technology and embrace instead a nuanced, broadly defined anthropological approach (Pfaffenberger 1992; Schiffer 2001). Within anthropology a classical approach to the coconstruction of objects, use, and users is the diachronic or narrative perspective embraced by Appadurai (1986) and others (see Woodward this volume). Viewing objects as markers of social relationships shapes not only the material properties of the objects themselves, but also the subjectivity of both objects and their users.

Together, all these processes show how concepts such as domestication take everyday contexts, users, uses, experiences, choices, and techniques as their points of departure. Thus both mundane and domestic contexts (e.g., Miller 2001b), as well as concepts and processes that are seemingly distant, immaterial, and even esoteric (at least in the popular imagination), are deeply anchored in mundane routines (e.g., Miller and Slater 2001). Approaching technological systems as the outcomes of experience and choice frees up space for the study of material technoculture as social practice (Suchman et al. 1999). Such focus on social practices opens new ground for processes—such as use and consumption—traditionally neglected at the expense of design and production (see, e.g., Cowan 1989; Holt 1995; Oudshoorn and Pinch 2003; Silverstone and Hirsch 1992; Woolgar 1991).

The Politics of Materiality

Let us return for a moment to my initial example of the technoculture of wine production and consumption. Let us hypothesize that after some time in the field a clear theme emerges from your findings. The theme has to do with the sweetness and fruitiness of wine taste. Sweet and fruity wines—you happen to find—are generally less valued than bold and dry wines. Suppose you also find that as a result these wines are marketed differently: sweet and fruity wines are marketed to women, whereas bold and dry wines are marketed to men. Soon enough you also find that they are produced differently, that the rituals and scripts surrounding their drinking are different, and that an entire schism along the lines of gender exists in the social world of wine-making and consumption. Sounds like a familiar theme, perhaps running across much of our culture as a whole? If it does it is because, as many scholars have remarked, objects are indeed political. Or, in other words, material objects work as resources for the establishment, maintenance, and regeneration of power and status differentials. Ranging from politics of taste and distinction (Bourdieu 1984), to the reification of socially constructed boundaries between the "natural" and "cultural" worlds (Haraway 1996), critical studies of material culture and technology consistently point to the direct implication of the material world—and simultaneously to its supposed neutrality, independence, and imperative self-sustaining logic of progress in the public opinion—in structures of domination (Winner 1978).

Gender is especially a growing theme in the literature, especially in the sociology of science and technology and within cultural studies (Lerman, Palmer Mohun, and Oldenziel 1997; Wajcman 1991, 2004). Feminist scholars have extensively criticized dominant gender hierarchies and binaries, and related notions of value in the material world, as well as the supposedly absent role of women in technological development. Women have been both underrepresented as innovators of technology (Cockburn 1985), relegated to the realm of domestic technological consumption (Cockburn and First-Dilic 1994; Pink 2004), and often duped by promises of technological innovation that were mistakenly hailed as harbingers of social progress and equality (Cowan 1983). Relatedly, the identity and subjectivity of women have been shaped by technological choices that reinforced existing binary gender oppositions and dominant sexist hierarchies—as vividly shown, for example, by Oudshoorn's (2003) fascinating study of the social construction of the birth control pill as a female, rather than male, resource and prerogative. Another important concept with regard to the study of everyday life is Cowan's (1989) notion of the consumption junction. Consumption junctions are the sites where consumers and users make choices between competing objects, techniques, and technical purposes. Choices express individuals' agency, their discretion, and values, thus evidencing that all users, regardless of gender and social position, are active participants in the technological process (Clarke and Montini 1993; Lerman, Palmer Mohun, and Oldenziel 1997; Star 1991).

Gender is not the only source of social inequality, of course. Material objects are known to play a pivotal role in the reproduction of power differentials in a wide variety of contexts. As Tilley (1991:149) has succinctly put it, not unlike texts, material objects "embody ideologies and powers which form an essential part of their nature." Thus material objects and technologies embody ideologies that serve regimes of colonialist and postcolonialist value (Myers 2001a, 2001b), property rights (Henare 2007; Rowlands 2002), the definition and contestation of landscapes (Bender 2002) and collective memory (Saunders 2002), and much more (see Rowlands 2005). Among the most notable interventions in this field is Langdon Winner's (1988) classic investigation of the architectural infrastructure of some areas of suburban New York. According to Winner, under the architectural direction of Robert Moses, access to wealthy neighborhoods was kept under control by the low clearance of bridges and overpasses, a clearance low enough to make it impossible for city buses and their supposedly less affluent riders to move in and out of places where their presence was undesirable. This, as Winner elegantly put it, showed that artifacts have politics (for a much subtler and acutely critical argument see Woolgar and Cooper 1999).

Another common theme with regard to the politics of materiality is the Barthesian idea of myth—a concept powerful enough to have informed much of the enterprise of cultural studies as a whole. Barthes's concept of myth, illustrated in his classic work *Mythologies* (1957), explains how texts and material objects naturalize regimes of values or ideologies. Naturalization works through the connotative properties of a sign. Any sign has both a literal denotation (the common sense interpretation of what the sign refers to), and a system of connotations that evokes ideas, beliefs, and values. A system of connotations is always a possible site for contestation, as various ideologies inevitably compete over how a sign should be interpreted and what its dominant meanings should be. Yet, in everyday life, selective connotations are often effortlessly neutered of their political and ideological value and presented as natural, universal, common sense, and normal interpretations of the meaning of signs. Through a variety of examples ranging from consumer items to popular culture rituals and texts Barthes showed how common functions of material objects and their ordinary meanings perpetuate dominant ideological positions and reinforce power differentials. Besides the idea of myth and the useful—though by now considered somewhat simplistic—distinction between the denotative and connotative properties of material objects, Barthes's work theorized material culture as a form of text to be deconstructed and posited the act and role of "reading" as pivotal within cultural reproduction processes, thus drawing near to each other interpretive practices from the humanities all the way to archaeology and communication and cultural studies (Tilley 1990).

In the end, whether one focuses on class, gender, status, or other sources of difference and forms of distinction, it is important to remark that material culture is indeed always political and never neutral. Yet, the ways in which artifacts manifest their value for polity formation, maintenance, and dissolution are diverse, complex, and nuanced. Contemporary perspectives in material culture and technoculture studies—as much of the social scientific landscape in general—have thus moved away from determinist or reductionist approaches to the politics of materiality. Thus, scholars have amended the ahistorical and apolitical trends of phenomenology, the simplistic hierarchies, the excessive textualism, and binary oppositions of structuralism, and the realistic tendencies of hermeneutics and interpretivism. In their respective places they have advanced instead a very diffuse—and somewhat diluted, perhaps—post-structuralist approach that describes and explains the politics of materiality in nonessentialist, nondeterminist, multivocal, contested, emergent, and relational ways (Olsen 2006).

Objectification

Let us conclude this introductory chapter briefly offering a panoramic review of the concept of objectification. Again, a brief example will serve the purpose of introducing the topic. Let us suppose that it is Christmastime and you receive a package in the mail containing a bottle of homemade wine. As you learn from reading the card that your gift came with, the wine was made for you by your old uncle from the countryside, the uncle who has not seen you in a few years, but still remembers your strong liking of white dessert wines. Pleased with the gift, and utterly surprised, you realize that you have not bought any Christmas gifts for your uncle this year, or even last year. Ready to make up for this, you rush to the computer, log onto the Internet, and manage to buy a pair of tickets to a concert taking place just a few miles from your uncle's home. Because the tickets will be shipped from the distributor directly to him, you decide to send in the mail a note to wish happy holiday and to thank him for taking the time to make and send you the wine. What does this example illustrate? It shows us that relationships, identities, values, and so forth take shape through material forms. This exemplifies the process of objectification.

If indeed the above is a good example of objectification, it should not take too long to realize that the concept of objectification more or less describes the subject matter of the entire field of archaeology, and much of anthropology, technology studies, and quite possibly more. The concept of objectification, however, has not received as much explicit interest as one might imagine. While excellent overviews and discussions are available (Miller 1987, 2005; Tilley 2006b), objectification remains for the most part one of those terms that people often invoke without a clear definition in mind, or understanding of what the idea involves. Because good explanations already exist, my scope in this section is to provide a brief and convenient summary of the idea, to explain how the concept can be further developed across the various disciplines and subdisciplines interested in the study of material culture and technology, and to refer the reader to a few examples in the research literature. So, let us begin with a better definition and summary of what is at stake.

As Miller (1987) explains, the idea of objectification finds its roots in the thought of Hegel. Simply put, for Hegel objectification refers to the process whereby human beings undertake a double process of separation first and reappropriation later of the material world. Thus, throughout our development we first realize that "we" are not the wine we drink: we are separate material entities with distinct sources of being. But in the process of separating ourselves from the material world outside of us we also create the condition whereby the world outside us becomes subject to our willful "manipulation"—to borrow a concept from George Herbert Mead—and thus our creative action. When understood this way, various kinds of objectifications lend themselves to our study as concrete embodiments of things other than themselves. For example, we could study a bottle of wine as isomorphic with a particular social condition or relationship. A good deal of material culture analysis, admittedly, begins from this basic notion: the notion that ideas, values, traits, beliefs, and so on come first and take material shape later. The idealism and dualism of this idea is obvious, however, and so is the analytical problem it generates: if materiality is in fact so epiphenomenal, why should we bother studying it instead of the very ideas, values, ideologies, and so forth that determine it? Unfortunately, a great deal of academic writing takes—willingly or unwittingly—this stance.

A truly material, as it were, approach to materiality and objectification is instead more complex and nuanced. As Tilley (2006b:60) writes, objectification is a profoundly dialectical and emergent process "implicated in action, in the physical production of things which are therefore active in the self-constitution of identities, and interactions between people" as well as throughout the social life phases of objects themselves (see also Kopytoff 1986; Miller 1987, 2001a). A more effectual approach to objectification, therefore,

> attempts to overcome the dualism in modern empiricist thought in which subjects and objects are regarded as utterly different and opposed entities, respectively human and non-human, living and inert, active and passive, and so on. Through making, using, exchanging, consuming, interacting, and living with things people make themselves in the process. The object world is thus absolutely central to an understanding of the identities of individual persons and societies. Or to put it another way, without the things—material culture—we could neither be ourselves nor know ourselves.

If, as Tilley rightly suggests, material culture and culture writ large are inseparable, so are material culture and technology. A technological vision of material culture (or a material vision of technology, if you like) reveals the dynamic, processual, active, creative role that both humans and nonhumans play in the constitution of their physical and social worlds. The process of objectification therefore helps to reveal how material forms are the media for the generation, reproduction, and transformation of sociality rather than epiphenomena of the mind or manifestations of language in a different form. Objectification is thus wholly reciprocal: "object and subject are indelibly conjoined in a dialectical relationship. They form part of each other while not collapsing into or being subsumed into the other. Subject and object are both the same, yet different" (Tilley 2006b:61).

Viewed this way objectification is a concept as much as it is something bigger: an umbrella category of material processes, an analytical approach, and even the foundations of a materialist paradigm. As such, it is also a convenient tool to subsume not only the other topics discussed in this paragraph, but also the various ways of knowing the material world discussed in part one of this book, and obviously the various methodological approaches discussed in the rest of this volume. Whether one focuses on objects as extensions of the self (Belk 1988), relationships (Valentine and Longstaff 1998), or the person (Layne 2000), on tools as the building bricks of society (Hughes 2005) or as elements for the configuration of the subjectivity of users of technology (Mackay et al. 2000; Woolgar 1991), on social contexts and relationships as the foundations for unique technocultures (MacKenzie and Wajcman 1999; Schiffer 2001), on global regimes of material value (Myers 2001a), on modernity (Miller 1994), or capitalism (Miller 1997) the process of objectification stands, undoubtedly, as the key link between materiality and techne, and among the disciplines and fields dedicated to the study of both.

References

Adler, Patricia, Peter Adler, and Andrea Fontana. 1987. "Everyday Life Sociology." *Annual Review of Sociology*, 13:217-235.
Appadurai, Arjun (Ed.). 1986. *The Social Life of Things*. Cambridge: Cambridge University Press.
Belk, Russell. 1988. "Possessions and the Extended Self." *Journal of Consumer Research*, 15:139-168.
Bender, Barbara. 2002. "Contested Landscapes: Medieval to Present Day." Pp. 141-174 in *The Material Culture Reader*, edited by Victor Buchli. London: Berg.
Bourdieu, Pierre. 1984. *Distinction*. New York: Routledge.
Buchli, Viktor. 2002a. "Introduction." Pp. 1-22 in *The Material Culture Reader*, edited by Viktor Buchli. New York: Berg.
—— (Ed.). 2002b. *The Material Culture Reader*. New York: Berg.
Casper, Monica and Adele Clarke. 1998. "Making the Pap Smear into the Right Tool for the Job." *Social Studies of Science*, 28:255-290.
Clarke, Adele and Teresa Montini. 1993. "The Many Faces of RU486: Tales of Situated Knowledges and Technological Contestations." *Science, Technology, and Human Values*, 1:42-78.
Cockburn, Cynthia. 1985. *Machinery of Dominance: Women, Men, and Technical Know-How*. London: Pluto.
Cockburn, Cynthia and Ruza First-Dilic. 1994. *Bringing Technology Home: Gender and Technology in a Changing Europe*. Milton Keynes: Open University Press.
Cowan, Ruth Schwartz. 1983. *More Work for Mother: The Ironies of Household Technology from the Open Hearth to the Microwave*. New York: Basic Books.
——. 1989. "The Consumption Junction: A Proposal for Research Strategies in the Sociology of Technology." Pp. 261-280 in *The Social Construction of Technological Systems*, edited by Wiebe Bijker, Thomas Hughes, and Trevor Pinch. Boston, MA: MIT Press.
Csikszentmihalyi, Mihaly and Eugene Rochberg-Halton. 1981. *The Meaning of Things: Domestic Symbols and the Self*. Cambridge: Cambridge University Press.
Dobres, Marcia-Anne. 2001. "Meaning in the Making: Agency and the Social Embodiment of Technology and Art." Pp. 47-76 in *Anthropological Perspectives on Technology*, edited by Michael Bryan Schiffer. Albuquerque: University of New Mexico Press.
Fischer, Claude. 1994. *America Calling: A Social History of the Telephone to 1940*. Berkeley: University of California Press.
Haraway, Donna. 1996. *Simians, Cyborgs, and Women*. New York: Free Association Books.
Heath, Christian, Hubert Knoblauch, and Paul Luff. 2000. "Technology and Social Interaction: The Emergence of Workplace Studies." *British Journal of Sociology*, 51:299-320.
Henare, Amiria. 2007. "Taonga Maori: Encompassing Rights and Property in New Zealand." Pp. 47-67 in *Thinking through Things: Theorising Artefacts Ethnographically*, edited by Amiria Henare. London: Routledge.
Holt, Douglas. 1995. "How Consumers Consume: A Typology of Consumption Practices." *Journal of Consumer Research*, 22:1-16.
Hughes, Thomas. 2005. *Human-Built World: How to Think About Technology and Culture*. Chicago: University of Chicago Press.
Ingold, Timothy. 1996. "Situating Action V: The History and Evolution of Bodily Skills." *Ecological Psychology*, 8:171-182.

———. 2000. *Perception of the Environment: Essays in Livelihood, Dwelling, and Skill.* London: Routledge.
———. 2007. "Earth, Sky, Wind, and Weather." *Journal of the Royal Anthropological Institute*, (N.S.) S19-S38.
Kopytoff, Igor. 1986. "The Cultural Biography of Things: Commoditization as Process." Pp. 64-91 in *The Social Life of Things: Commodities in Cultural Perspective*, edited by Arjun Appadurai. New York: Cambridge University Press.
Layne, Linda. 2000. "He Was a Real Baby with Baby Things: A Material Culture Analysis of Personhood, Parenthood, and Pregnancy Loss." *Journal of Material Culture*, 5:321-345.
Lemonnier, Pierre (Ed.). 2002. *Technological Choices: Transformation in Material Culture since the Neolithic.* London: Routledge.
Lerman, Nina, Arwen Palmer Mohun, and Ruth Oldenziel. 1997. "Versatile Tools: Gender Analysis and the History of Technology." *Technology and Culture*, 38:1-8.
Lie, Merete and Knut Sorensen (Eds.). 1996. *Making Technology Our Own? Domesticating Technology into Everyday Life.* Oslo: Scandinavian University Press.
Mackay, Hugh, Chris Carne, Paul Benyon-Davies, and Doug Tudhope. 2000. "Reconfiguring the User: Using Rapid Application Development." *Social Studies of Science*, 30:737-759.
MacKenzie, Donald and Judy Wajcman (Eds.). 1999. *The Social Shaping of Technology.* Boston, MA: McGraw-Hill.
Martin, Michele. 1991. *"Hello Central": Gender, Technology, and the Culture in the Formation of Telephone Systems.* Montreal: McGill-Queens University Press.
Miller, Daniel. 1987. *Material Culture and Mass Consumption.* New York: Blackwell.
———. 1994. *Modernity: An Ethnographic Approach.* London: Berg.
———. 1997. *Capitalism: An Ethnographic Approach.* London: Berg.
——— (Ed.). 1998. *Material Culture: Why Some Things Matter.* Chicago: University of Chicago Press.
———. 2001a. "Alienable Gifts and Inalienable Commodities." Pp. 65-90 in *The Empire of Things*, edited by Fred Myers. Santa Fe: School of American Research Press.
——— (Ed.). 2001b. *Home Possessions.* London: Berg.
———. 2005. "Materiality: An Introduction." Pp. 1-50 in *Materiality*, edited by Daniel Miller. Durham, NC: Duke University Press.
Miller, Daniel and Don Slater. 2001. *The Internet: An Ethnographic Approach.* London: Berg.
Miller, Daniel and Chris Tilley. 1996. "Editorial." *Journal of Material Culture*, 1:5-14.
Moser, Ingunn. 2000. "Against Normalisation: Subverting Norms of Ability and Disability." *Science and Culture*, 9:201-240.
Myers, Fred (Ed.). 2001a. *The Empire of Things.* Santa Fe: School of American Research Press.
———. 2001b. "Introduction: The Empire of Things." Pp. 3-64 in *The Empire of Things*, edited by Fred Myers. Santa Fe: School of American Research Press.
Olsen, Bjørnar. 2006. "Scenes from a Troubled Engagement: Post-structuralism and Material Culture Studies." Pp. 85-103 in *Handbook of Material Culture*, edited by Chris Tilley et al. London: Sage.
Oudshoorn, Nelly. 2003. *The Male Pill: A Biography of a Technology in the Making.* Durham, NC: Duke University Press.
Oudshoorn, Nelly and Trevor Pinch (Eds.). 2003. *How Users Matter: The Co-construction of Users and Technology.* Boston, MA: MIT Press.
Pfaffenberger, Bryan. 1988. "Fetishized Objects and Humanized Nature: Towards an

Anthropology of Technology." *Man*, 23:236-252.

———. 1992. "Social Anthropology of Technology." *Annual Review of Anthropology*, 21:491-516.

Pink, Sarah. 2004. *Home Truths: Gender, Domestic Objects, and Everyday Life.* London: Berg.

Rowlands, Michael. 2002. "The Power of Origins: Questions of Cultural Rights." Pp. 115-134 in *The Material Culture Reader*, edited by Victor Buchli. London: Berg.

———. 2005. "A Materialist Approach to Materiality." Pp. 72-87 in *Materiality*, edited by Daniel Miller. Durham, NC: Duke University Press.

Saunders, Nicholas. 2002. "Bodies of Metal: Shells of Memory: 'Trench Art' and the Great War Re-cycled." Pp. 181-206 in *The Material Culture Reader*, edited by Victor Buchli. London: Berg.

Schiffer, Michael Bryan (Ed.). 2001. *Anthropological Perspectives on Technology.* Albuquerque: University of New Mexico Press.

Sillar, Bill. 1996. "The Dead and the Drying: Techniques for Transforming People and Things in the Andes." *Journal of Material Culture*, 1:259-289

Silverstone, Roger and Eric Hirsch (Eds.). 1992. *Consuming Technologies: Media and Information in Domestic Spaces.* London: Routledge.

Star, Susan Leigh. 1991. "Power, Technology, and the Phenomenology of Conventions: On Being Allergic to Onions." Pp. 26-56 in *A Sociology of Monsters: Essays on Power, Technology, and Domination*, edited by John Law. New York: Routledge.

Suchman, Lucy, Jeanette Blomberg, Julian Orr, and Randall Trigg. 1999. "Reconstructing Technologies as Social Practice." *American Behavioral Scientist*, 43:392-408.

Tilley, Chris. 1990. *Reading Material Culture: Structuralism, Hermeneutics, and Post-structuralism.* London: Basil Blackwell.

———. 1991. *Material Culture and Text: The Art of Ambiguity.* London: Routledge.

———. 1997. *A Phenomenology of Landscape: Places, Paths, and Monuments.* London: Berg.

———. 2006a. "Identity, Place, Landscape, and Heritage." *Journal of Material Culture*, 11:7-32.

———. 2006b. "Objectification." Pp. 60-73 in *Handbook of Material Culture*, edited by Chris Tilley. London: Sage.

Tilley, Chris, Webb Keane, Susanne Küchler, Mike Rowlands, and Patricia Spyer (Eds.). 2006. *Handbook of Material Culture.* London: Sage.

Valentine, Gill and Beth Longstaff. 1998. "Doing Porridge: Food and Social Relations in a Male Prison." *Journal of Material Culture*, 3:131-152.

Wajcman, Judy. 1991. *Feminism Confronts Technology.* New York: Polity Press.

———. 2004. *TechnoFeminism.* New York: Polity Press.

Winner, Langdon. 1978. *Autonomous Technology: Technics-Out-of-Control as a Theme in Political Thought.* Boston, MA: MIT Press.

———. 1988. *The Whale and the Reactor: A Search for Limits in an Age of High Technology.* Chicago: University of Chicago Press.

Woolgar, Steve. 1991. "Configuring the User: The Case of Usability Trials." Pp. 57-102 in *A Sociology of Monsters: Essays on Power, Technology, and Domination*, edited by John Law. New York: Routledge.

Woolgar, Steve and Geoff Cooper. 1999. "Do Artefacts Have Ambivalence? Moses' Bridges, Winner's Bridges, and Other Urban Legends in S&TS." *Social Studies of Science*, 29: 433-449.

2
Actor-Network Theory: Translation as Material Culture

Grant Kien

Writing is an act of poaching: stealing phrases, words, scenes, and experiences from the world around oneself, rearranging them, and in so doing, claiming selected bits for oneself as an author. The performance of writing makes concept into material—in the materiality of process, if no longer in the materiality of print. The act of reading is writing's second material moment. Partial, arbitrary, strategic, writing is translation: it is a struggle for meaning, not necessarily the "correct" meaning, but rather the will to be meaningful and communicate with recognized authority. This is a version of a story of the ways I have encountered Actor-Network Theory (ANT) and how I translate those encounters to explain material culture. The body of work comprising ANT spans more than three decades, generated mainly by Michel Callon, Bruno Latour, John Law, and their associates (with many important authors and works left for you to explore on your own should this brief introduction perk your interest). The term *relational materiality* (Law 1999) is of utmost importance to this writing, as is the notion of the *immutable mobile* (Law 1986) and the concepts of *inscription* and *durability* (Latour 1988a). Before I proceed, let me clarify that the terms technology, artifact, token, and material object will be used somewhat interchangeably in this chapter, unless stated otherwise.

Now, to be fair, it was never Actor-Network theorists' goal to explain material culture. They in fact dedicated themselves to transcending modernist paradigms that split the world into discrete conceptual realms set against one another, such as material versus metaphysical. Rather, as a way of getting past such problematics, they chose for the most part to ignore them, concerning themselves more with a way of mapping relationships between specific actors and actants that give rise to phenomena, and speculating on the nature of those relationships. The patterning of such relationships—the ordering of the world through ritual as repetitive, communicative acts—requires another sort of conversation to be had before getting into the question of how to map and the role of materiality in ANT. For this, I look to James Carey's (1989) work on communication as culture. Ideas of time are also important. Time structures the every day (Heidegger 1996), and it is this structuring of the days, weeks, months, and years through repetition that is the genesis of ritual (Carey 1989). We cannot live the same moment twice, but we *can* act out the same fictionalized moment ad infinitum. I suggest that it is what we do with these rituals that give rise to the phenomenon of material culture, making material meaningful.

The Material of Culture

Embedding regimes of practice into material objects helps ensure constancy across time, just as inscribing practices of production into technologies such as robots and other machines ensures consistency in material goods. Take pyramids, for example. Why did the first peoples of Mexico feel it necessary to make their culture materialize in the form of giant stone pyramids? Such tremendous sacrifice and effort is what James Hay (2001) would describe as *mattering*, or creating media to make both material artifact and symbolic meaning at the same time, programming a regime of behavior into the materiality of the phenomenon. With time, the Mexican pyramids have become repurposed. The rituals in which they participate have changed from sacredly religious to sacredly consumerist. They remain at the center of ordering meaningful ritualized performances of culture nonetheless. One may hypothesize that many of our motivations in the present are not so much different from the ancient Maya—to stabilize the relations that make our everyday lives meaningful through routinized, ritualistic performativity.

Material goods can be understood to both work symbolically and to be enlisted as allies to communicate belonging/sameness, otherness, and to stabilize everyday life. Lury (1996) wrote that the term material culture is used to describe "person-thing" relationships, and those things in use. Material and culture are always related, but material culture is concerned with the conversion of material objects toward a personally experienced end use. The utility of goods is always framed by cultural logic, and goods are imbued with significance through ritual. Material objects thus make visible the categories that we use to conceptually organize the world in which we live. Totemism, for example, binds tribal people together. Material objects both work and are worked upon, and in the process gain importance through the histories associated with them. Cultural significance is then gained through "paths and diversions through which an object circulates, accumulating a life history of associations and meanings" (22). This life history of objects provides a seeming depth to the otherwise purely aesthetical phenomenon of the performative use of artifacts.

Rituals hold individuals in a stable set of relationships with others and the world they live in (Carey 1989). Symbolic form (abstract signifiers) and narrative (myths that define and give meaning to the symbolic) are the conceptual aspects of ritual, but it is in "danced ritual" that we can identify the enactment of spatial form, and with it the materiality of culture. The idea of danced ritual directs our thinking of material culture to process rather than form. Without conceptual information and the processes of making objects meaningful, material objects fail to "matter" and become deadweight to be cast off rather than important elements of everyday life. ANT is thus, among other things, a method of mapping how a technology/artifact/material object participates in everyday life.

A Sociology of Translation

ANT first appeared in the writing of Callon in 1980 to describe the actions of engineers whose work entails creating what would come to be known as "messy networks" (Bijker and Law 1992:12) of technology, society, and economics. The obvious product of engineering is the material thing brought about by the design and production process, but Callon demonstrated how engineers are themselves constituted and shaped in those same productive networks. Callon had effectively shown that the neutrality of modernist science was a myth, and the process itself was anything but linear or purely logical in nature. Soon after, Callon and Latour (1981) introduced one of the main ANT premises: the rejection of both dualism and of the idea that macro and micro phenomena are essentially different or apart. Rather, they proposed an early theory of translation to describe the way that the Hobbesian leviathan is made manifest as a seeming entity through the ongoing work of many actors. They proposed that any social phenomena can be analyzed using the model of translation, which shows categorical distinctions and dualisms to be theoretical, not essential. Callon outlined three methodological principles for the "sociology of translation":

1. The observer should be agnostic (that is, should believe foundations and absolutes are not possible) and refrain from judgments and interpretations.
2. Conflicting viewpoints and arguments in any controversy should be explained in the same terms to maintain "generalized symmetry" and thus neither technical nor social aspects should entail changes in the language of translation.
3. "Free association" means leaving aside distinctions between "natural" and "social" phenomena, rejecting a priori categorization to allow actors to define and associate the elements of their world according to their own language (200–201).

Callon's goal was to bypass the chicken/egg paradox of dualist thinking and privileged hierarchies of knowledge. He asserted that the voices of the subjects should be studied as valid sources, affirming that this method allows the nuances and contradictions in the evolution of networks of relationships to be better understood (201). Such understandings are not essential truths themselves: "translation is a process before it is a result" (224). As practice, translation projects a deceptive appearance of essential definitions, but such distinctions are in reality never so definitive or distinct as the translation of them would have us believe. Be this as it may, Callon wrote (1986:224), "Translation is the mechanism by which the social and natural worlds progressively take form. The result is a situation in which certain entities control others." In this quote, we see the fruition of the path from concept to materiality expressed in word form, and the critical concern with power. Latour later schematized the process as a way of understanding power relations.

Translation and Power

Is nature powerful? Is a gun or car, or the president of the United States? Does money make people powerful? Latour stated that in working with a model of power as translation, power is the effect of performance and not something in possession of the performer (actor). The effect (i.e., power) is produced by associating entities together. Power involves a paradox: having power means having unused potential (i.e., nothing), while using power means other people or some object are performing for your benefit. Power is the performance of acts by others, or the potential of having others act for oneself. So, a translation model of power rejects objectifying power. Rather, translation theory considers objects and objectified conceptual symbols as fetishized and reified as containers of power signifying power in itself. Latour (1986) describes three elements important to the spread of such "tokens" (artifacts, claims, orders, etc.) of power in time and space:

1. The only energy of the token is the force that instigates the movement and sends it on its way.
2. A token travels with inertia (as in Newton's law), thus it will travel freely as long as there is a lack of opposing forces to halt the movement of the token.
3. The medium the token travels through is a network of actors that reshape and transform the token (translate it) as they pass it along (266-268).

Latour's explanation of power thus moves attention to everyday practices in which power tokens are enacted and thereby passed along, such as the transactions that pass money from one actor to another, or the actual circumstances surrounding the shooting of a bullet. A bullet, for example, continues along its path as far as opposing forces such as gravity will let it, and it is transformed by the media through which it passes, such as the grooves of a gun barrel. No one can ever claim power as property: "'Power' is always the illusion people get when they are obeyed" (268). Only by giving up the effort to explain the origin of power and instead work forward from the assumption that it exists can we begin to study it. For Latour, power is thus a "consequence," not an origin (271), the material aspect of which is part of a process, and thus he and other translation theorists turned their attention to society as the performance of relationships. And here we have the ethnographic imperative.

Like Callon, Latour took issue with traditional (Durkheimian) ethnomethodology that continues to try to separate "society" from that which composes it and makes it durable ("social and nonsocial elements") (275). A traditional approach works as if society were some kind of glue with bonding power. Latour was firm in his criticism of this approach: "society is not what holds us together, it is what is held together. Social scientists have mistaken the effect for the cause, the passive for the active, what is glued for the glue" (276). He stated that society can be transformed from something that is a principle (unknowable) into something knowingly practiced. Studying the techniques of performance is part of this knowledge. Among the assets of this approach, a performative definition of society tells us:

1. A definitive list of the "properties" of social life (principles) is theoretically impossible, although it is possible in practice (by simply listing what really is practiced).
2. Society (the whole and the parts) is defined by the practices of actors.
3. An actor's quantity of knowledge is not important, rather their definitions of how they contribute to an overall picture is.
4. Actors and social scientists are interested in the same questions but have different practical ways to enforce their definitions of society (273).

Latour concludes that "the notion of power should be abandoned," with attention turned instead to "the stuff of which society is made" (278).

The Stuff of Society and Nonhuman Actors

From their early work on, Latour and his peers (particularly Woolgar) engaged with examining the production of scientific knowledge (see Latour and Woolgar 1979), helping found the field of the sociology of science. They were very successful in illustrating that the production of scientific facts relies heavily on social forces and politicized contingencies. They introduced the notion that the objects of study were also noteworthy participants in the fact-making process (Law 1986). Using a case study of fifteenth- and sixteenth-century Portuguese trade, Law described how Latour's concept of the immutable mobile was enacted in Portugal's global trade networks. That is, the ships themselves were integral to the successes and failures of Portugal's trade networks, which entails accounting for the research and development of naval advances and shipbuilding along with the economics and politics that made such advances possible. The ships exemplify the ability of some tokens to travel great distances and endure numerous translations, yet retain much of their original form. Here we come to a key feature of the nonhuman actor.

Latour (1988c) explained that machines are called on as political allies just as humans are. This assertion builds on the aforementioned premise that the technology/society distinction is the result of a misleading theoretical fabrication that leads one to think of technology as nonpolitical, making it mistakenly appear that the power of technology is the power of humans. In reality, machines are part of the political ordering of the world, employed as very sophisticated allies in the recruitment of other allies and keeping them in place. The inertia of technology becomes a materialized illusion that reifies the dominating actor as possessing their power. Latour used the historical development of the Parisian subway as an example: the maintenance of positionality through managerialism in the Parisian subway/railway fight led to narrower tunnels than the railway could use in order to maintain the positions of the subway managers and keep their agency from being swallowed by the national railway. Over time, this managerial decision became naturalized in engineering practice, until it was later reversed and the tunnels were made larger. However, in the interim, the tunnel design and subway/railway split had become part of the cultural fabric of Paris. This example illustrates the enlistment of seemingly benign and mundane material objects for political maintenance, but it also introduces an important new term: the ally.

Latour posited that authority figures work as part of "administrative machinery" and actually resist individuation in the practice of translation. As he wrote,

> it is useless to impose a priori divisions between which skills are human and which ones are not human, which characters are personified and which remain abstract, which delegation is forbidden and which is permissible, which type of delegation is stronger or more durable than the other (1988a:305).

He described translation as "the translation of any script from one repertoire to a more durable one" (306). By "repertoire," he meant translating a performative script from a human body to a machine, or vice versa. This entails *inscription* or *encoding*. "Prescription" is whatever is presupposed from "transcribed actors and authors," including presuppositions encoded in machines (306) (e.g., "inscribed readers" can be generated by a process similar to Althusser's description of interpellation). The inference is, "des-inscription" is breaking from prescribed behavior, while "subscription" is acquiescing to it (307). Turning to context, "pre-inscription" is everything that prepares the scene for articulation (307). More directly related to the actor, "sociologism" is the ability of one to read the scripts of nonhuman actors, and "technologism" is the ability of humans to read their own behavioral scripts prescribed by the technology (307-308).

Building on Marx's premise that technology reliably replaces unreliable human activity, the performative aspect of machines is rather straightforward: "Machines are lieutenants; they hold the places and the roles delegated to them" (308-309). Along with the functional aspect of a machine's encoding, cultural values are also prescribed: "What defines our social relations is, for the most part, prescribed back to us by nonhumans" (310). Machines thus perform back to humans the cultural values the original human actor that they replace was meant to stabilize (his example is that of the door closer, which performs the act of good manners in opening doors for people). Thus, "studying social relations without the nonhumans is impossible" (310), and herein is the intersection with our fascination with technology and materiality directly inspiring the book in your hands. As Vannini writes in the introduction, the time is at hand to decenter mainstream ethnography by paying as much attention to techne as ethnos. This central premise of the ethnography of technology and material culture, and ANT, not only allows some important ontological implications to be elaborated, but also moves from a modernist obsession with causality to a concern with effects reflected in the latest advances in ethnographic methods.

How Nonhumans Act

The most descriptive theoretical writing in the ANT canon is "Irreductions," published as the second part of the volume containing Latour's famous study *The Pasteurization of France* (1988b). This writing itemizes point by point the rhetorical and philosophical assumptions that order actors, networks, and the process of translation together into the appearance of a theoretical system and applicable method. In it Latour begins by clarifying the term actor:

> I use "actor," "agent," or "actant" without making any assumptions about who they may be and what properties they are endowed with. ...they have the key feature of being autonomous figures. Apart from this, they can be anything—

individual ("Peter") or collective ("the crowd"), figurative (anthropomorphic or zoomorphic) or nonfigurative ("fate") (252n11).

The aesthetic of autonomy defines the actor/agent/actant, but the issue of agency (a.k.a. autonomous power) must then be dealt with. In keeping with the law of inertia, Latour suggests, "In place of 'force' we may talk of 'weaknesses', 'entelechies,' 'monads,' or more simply 'actants'" (Latour 1988b:159). The actant thus becomes the primary unit in building networks, assembling relationships through the power of inertia and the weakness of other actants to halt its monadic[1] drive for self-fulfillment. Any actant can do this: "No actant is so weak that it cannot enlist another. Then the two join together and become one for a third actant, which they can therefore move more easily. An eddy is formed, and it grows by becoming many others" (159). Actants must be enlisted (i.e., persuaded) into the network. To maintain its position, an actant must constantly build relationships with other actants. The more relationships, the more stable (autonomous) an actant appears to be. Strength is gained by association, by speaking on behalf of all other actants in the network, and in effect, by translating the voices of the multitude of other actants. But the other actants do not simply fall silent; they may in fact be silenced, fall mute of their own power, or have individual voices lost in the cacophony of the crowd. The dominating voice justifies itself as democratic, enunciating what the network demands of it, which is in effect quite true, since the dominating voice will only say that which creates new alliances and maintains the network.

The only effective oppositional force to network can be a rearranged array of alliances. This irreversible struggle is asymmetrical, meaning there are clear winners and losers. Actants simply "lean" on a more "durable" force to create the appearance of being "more real" to other actants, and thus win their acquiescence (160). Laws or structures are learned by one actant from others, until the student begins to enact its own version and make other actants join it.

As mentioned, one of the key features of translation as an organizing process is the abolition of dualistic categories, such as sameness and difference. Thus, "nothing is, by itself, the same as or different from anything else. That is, there are no equivalents, only translations" (162). Identification between actants is laboriously constructed from "bits and pieces," and vigorously maintained. Exchanges are always unequal, thus costing actants to create and continue relations. This "principle of relativity" claims that "the best that can be done between actants is to translate the one into the other" (162). Everything is relative and unequivalent, in contrast with the modernist approach that seeks to make the observer greater than or equivalent to the observed. Being relative, every alliance/relationship is negotiable.

While a network is a spatialization, actants themselves can neither be totaled nor definitively located. Rather, location is "a primordial struggle" in which many actants get lost (164). Contrasted with static mapping, location is a process of an actant finding others, or being found. Thus time and not space is what is contested. But actants are not framed by time and space (which would reify modernist dualism), rather time and space become descriptive frameworks only for actants temporarily under the hegemonic influence of another. Method here becomes an issue.

ANT as Methodology

If time and space are hegemonic impositions, how then is one to write? There is only translation, not knowledge, and even morals are provisional. As Latour (1988b:167) instructs, "we should try to write a text that does not take time and space but provides it instead." Misunderstanding and inaccuracy is inevitable, but the more active one is, the more one is able to negotiate and/or impose one's own definitions and measures. An actant increases its appearance of strength and influence by persuading or coercing other actants to fall into its own network array. Stronger than the others, it "translates, explains, understands, controls, buys, decides, convinces, and makes them work" (172).

Translation is a process of narrative construction, one explanation following and flowing from another. But it is never exact; since one language cannot be simply reduced to another translation, it is always a hegemonic practice, but it is a practice that every actant understands: "daily practice needs no theorist to reveal its 'underlying structure'" (179). It is not the "best" translation that wins out, rather the seemingly most powerful. Messages change and mutate through transport and translation, but languages themselves are also forces, actants being words that might also be things. A sentence whose alliances are intact appears to be true, while a sentence with insufficient allies appears false. Truth and falsehood lose their usefulness as foundational premises (188). Rather than seeking evidence of truth, Latour suggests a more appropriate method: "to be profound, we have to follow forces in their conspiracies and translations...wherever they may go, and list their allies, however numerous and vulgar these may be" (188).

Individual actants each construct their own complex world, always in negotiation, sometimes the hegemonic force, sometimes subjected to hegemony, sometimes the translator, sometimes silent. However, every actant in the network participates somehow in deciding which actant will speak and when. An actor can only prove its power to the extent that other actants in turn say for themselves they want the same thing, and that is thus the extent to which an alliance is reliable. Some actants gain the temporary power to speak on behalf of the multitude and thereby define other actants, linking them together in a network that only the temporarily dominant one can translate. However, no acquiescence is ever complete.

No actant ever stops acting in its own self-motivated interest, even when it willingly joins or is coerced into a strong alliance. Since the dominant actant can only be stopped by the morphing or extension of networks into other networks, there are always competing interests that seek to divert the network of alliances toward their own ends. Thus, faithful allies make stronger entelechies, less easily diverted. Potency is an aesthetic spatialization of setting forces against each other, collecting the faithful as inside pushing against the doubtful (outside) as in the reification of modernist essentialisms. However, in practice it is not so simple as maintaining imagined borders. Actants try to program other actants to go against their nature and not betray the dominant ordering, extending the dominating actant and making its array of alliances a performed hierarchical reality.

Latour punctuated his elaboration of actors and networks with what strangely sounds like a universal: given that there is no modernism, all networks can be described the same way. However, he explained this statement in terms of a process of translation, as a translator himself:

> Since there are not two ways of knowing but only one, there are not, on the one hand, those who bow to the force of an argument, and on the other, those who understand only violence. Demonstrations are always of force, and the lines of force are always a measure of reality, its only measure. We never bow to reason, but rather to force (233).

The work of Callon, Latour, and Law stand out as apexes in the sociology of translation that emerged in the 1980s. Law in particular found this approach useful for managerial and organizational studies. Actor-Network studies of science and technology also continued, but early in the 1990s the voice of a symbolic interactionist pointed out a problematic recurrence in ANT literature.

The Problem of Singularizing Identity

In symbolic interactionist fashion in 1991, Star pointed out that it is easier for some to perform translation, arguing that the experiences of the marginal provide opportunities to understand that identity is never singular, and is sometimes built on a resistance to translation itself. Not everyone is translatable, in other words. In practice, it is therefore the performance of multiple identities that enables control of power networks. Power in networks depends on "processes of delegation and discipline" (Star 1991:28), but no one human actant is *always* identified the same way. One who is a manager for specific hours of the day may also be a parent during other parts of the day, a subservient another time, a spouse, and so on. More poignant is the example of those who are marginal and kept silent in society, due to practices of various isms (racism, sexism, ableism, etc.). Star claimed that multiple personality and marginality are important points of departure in feminist and interactionist studies of technology and power. Multiplicity is the result of being constantly delegated and disciplined, in contrast with those who delegate and discipline. Such a multiple self is "unified only through action, work and the patchwork of collective biography" (28). For people in such positions, the self is constructed and maintained in refusing to accept the attribution of one's efforts to an executive, refusing to give up the multiplicity of the hybrid self and become "pure," and acknowledge the multiple worlds in which marginality has membership.

Star agreed that networks are made stable with standards and conventions, which are revealed in transgressions and nonconformities. Something as simple as customizing a fast food order can reveal the performance of marginalization as the system cannot handle originality, rather requires a critical mass to influence change in the whole. She pointed out that

> [p]eople inhabit many different domains at once... It's important not to presume either unity or single membership, either in the mingling of humans and nonhumans or amongst humans. Marginality is a powerful experience. And we are all marginal in some regard, as members of more than one community of practice (social world) (52).

Of course, marginality is only powerful when power is granted to it, which is a process Law explicated in later writings. However, given the multiplicity of the marginal, Star suggested a useful place to begin to understand technology, and power is from the point of transgression and contradiction where multiplicity is at odds with itself, revealing hegemony in practice.

Law (1992:380) seemingly addressed Star's concern, explaining ANT's attention to "the mechanics of power" by emphasizing how "society, organizations, agents and machines are all *effects* generated in patterned networks of diverse (not simply human) materials" (380, original emphasis). He claimed that "network consolidation" (380) is only the appearance of singularity of a network. Since networks are ordered with materials and strategies, patterns generating effects of power and hierarchy in institutions and organizations, ANT suggests that "the social is nothing other than patterned networks of heterogeneous materials" (381) composed of people, machines animals, texts, architecture, symbolic material (money), and so on. Hence, "the task of sociology is to characterize these networks in their heterogeneity, and explore how it is that they come to be patterned to generate effects like organizations, inequality, and power" (381). This statement fits nicely with the more recent turn in ethnography to performance and its concern with effects rather than causes. The key point here is that what we consider a person is an "effect" produced by "a network of heterogeneous, interacting, materials" (383).

Locating Agency and Things

The body is *not* where either agency or the actor is located. In keeping with ANT's theory of power, actors are found in network patterns and "heterogeneous relations" (Law 1992:384). An actor is the temporal sum total of all the relations that compose it. The effects mask the networks that produce them (e.g., a working television, a healthy body), and we tend to deal with the effects of networks rather than an "endless network of ramification" (385). When a network pattern is "widely performed" (385) it becomes a thing (an alliance) one can count on as a resource, and hence serves as a network interface. Making an aesthetic structure is a process, "a site of struggle, a relational effect that recursively generates and reproduces itself" (386). There is never completion, autonomy, or real singularity. Exposing the process of how actors create the appearance of singularity and autonomy in spite of the reality of multiplicity and relativism are core concerns of ANT, a concern with power as effect, not causality.

Law described ANT's method as telling "empirical stories about processes of translation" (387) in so doing analyzing tactics and strategies. A good strategy embodies relations in "durable materials" (387) to maintain network stability through time, and materiality is thus always strategic. But durability is itself an interactive effect, lest the material object lose any signifying power and simply drop out of the network. Surveillance and control produce the effects of centers and peripheries, while translations create immutable mobiles (387). "Centers of translation" (the anticipation of response and reactions of those being translated) are relational effects, generated by conditions and material that can also "dissolve" them (388).

Translation is local, but there are general strategies of translation that reproduce in ranges of networks. Numerous strategies may be manifest at the same time, may constitute what is called an "organization," or may generate network stability, durability, mobility, hierarchies, asymmetries, and so on. Law concluded his chapter with a statement on how he hoped to contribute to the age-old Marxist dilemma of reproduction: "To the extent that 'society' recursively reproduces itself it does so because it is materially heterogeneous. And sociologies that do not take machines and architectures as seriously as they do people will never solve the problem of reproduction" (389). Law's comment in fact appeared around the same time frame as the emergence of critical geography and authors such as Harvey, Lefebve, Soja, and Foucault, who fabricated a discourse seeking to understand the structuration of ideology in the material artifacts of society.

In 1995 the first work that seemed to specifically address Star's criticism appeared. Callon and Law (1995:483) firmly stated that ANT needed to transcend the "egalitarian panopticism of liberalism." They argued that it cannot be proven whether humans and nonhumans are alike or not in terms of agency. They suggested that leaving the question unanswered (can nonhumans have agency?) is the most interesting way to proceed (483). They reaffirmed Law's previous relational ontology argument that there are no things in and of themselves, introducing the new term "collectif" to explain agency.

Callon and Law described agency as the result of translating for a *hybrid collectif*, an array of relations, links, interpenetrations, and processes (this can be contrasted with "a collective," which is a thing). Not all collectifs can be agents, but some are. A collectif is not personal, since it is not dualistic (no outside/inside, not extrinsic), but rather includes all that inspires, influences, and touches it. In this understanding of agency, differences and dualism is generated out of partial similarities. Here there is a distinction to be made between "actant-network" and "Actor-Network": "All actants are created equal. But actors have distributions thrust upon them" (490). The distinction is one of function and will, actants simply performing their inscribed roles, while actors struggle with ill-fitting definitions. The collectif allocates agency to a particular area, creating specific classes of agents and the actants in those places are then *said* to be agents. Every array of agency is different, but translation is used to "build readers and worlds all together, at the same time" (500). The chains of translation that construct agents cannot be easily distinguished in a macro perspective, but can be understood on a local, purposeful level.

Signification happens in myriad ways, including in hybrid forms, not just in language alone. Nondiscursive signification also represents, and nonhuman actors also order the world, but there is a bias toward the human translation because of the bias for language. There are political implications of these localized appearances of agency that seem to happen in the form of the human body. Such allocated spaces are not a homogeneously "liberal," but a set of overlapping places that can only be partially imagined and not assimilated. The error is in thinking that all "others" can be assimilated. Rather, many "others" will simply ignore the networks that seek to turn their performances toward other ends. What is particularly interesting for me here is in thinking through Star's suggestion that the marginal find definition in resistance, for in resistance such actors are profoundly joined with the hegemony. Callon and Law go a step further in showing how actors can in practice exist not outside of, but rather in unsanctioned places in spite of Actor-Networks.

Latour (1998) pointed out that ANT had never made the claims that Star criticized, emphasizing that ontologically ANT is unashamedly reductionist and relativist in saying there is nothing but networks with no space between them, as "a step towards an irreductionist and relationist ontology." Adding ontology (actor) modifies the mathematical construct of the network, such that spatializations (close/far, up/down, local/global, inside/outside) are "replaced by associations and connections" (ibid.). Sketching the properties of Actor-Networks "moves on from static and topological properties to dynamic and ontological ones" (ibid.). Latour clarified that ANT is not a theory of action, but rather serves like a map for navigators, providing suppositions of what one might encounter in a given journey. Networks create their own frames of reference, which accounts for the postmodern concern with reflexivity.

ANT Translates Itself

By 1999, ANT was a very successful perspective. The publication of *Actor Network Theory and After* (Law 1999) clarified some of the nagging misunderstandings and misapplications of the canon, and some of the personal misgivings of its main voices. Law labeled ANT a "semiotics of materiality" (4). He used the term "relational materiality" (4) to describe how form is an effect of relations, made durable and fixed through performance. He suggested that ANT is always in tension between agency (actor) and structure (network). He also described that explanations of network assemblage are prone to Machiavellian and managerial answers, and discourses of strategy, but this is problematic because it only explains the struggle to a center from a center. It does not translate the experience of being marginal, does not account for nonstrategic order, does not explain that which is inassimilable, and overall ignores hierarchies. Spatialization as network achieves object integrity by keeping patterned links stable, constituting regions with networks (i.e., nations are made of numerous networks superimposed) (6–7). So by fighting against essentialisms, ANT reveals that arbitrary orderings can be otherwise. It has helped destabilize Euclidean (flat) geometry by showing that what is taken to be "natural" form is in fact produced in spatial networks (8).

Latour took issue with the semantics of ANT, pointing out the word network was meant to provide the language of social theory a signification of "series of transformations—translations, transductions" (15), a concept of constant mutation, not the static formation it is now popularly taken to signify. The coupling of actor with network unfortunately gives the appearance of a fascination with agency and structural dichotomies, micro versus macro, and so on, whereas the goal is rather to bypass and ignore the issue of opposition by "summing up interactions through various kinds of devices, inscriptions, forms and formulae, into a very local, very practical, very tiny locus" (17). ANT does not seek to describe how actants act, but rather, "what provides actants with their actions, with their subjectivity, with their intentionality, with their morality" (18). ANT researchers do not "study" actors, but rather connect with them, recording how worlds are built in specific sites.

Callon continued the critical analysis of the actor: "ANT is based on no stable theory of the actor; rather it assumes the *radical indeterminacy* of the actor" (181). Size, psychology, and motivation are not considered predetermined, opening the social to nonhumans, overcoming individual/holistic dichotomy, and demonstrating language as an effect rather than absolute determination. The identification of actors and actions depends on the configurations of the networks they appear in, and can only be understood "if we agree to give humans all the nonhumans which extend their action" (194).

Post-ANT

The publication of *Actor Network Theory and After* punctuated a moment of maturity in the perspective, but it did not signal the end for ANT. In 2000, Law clarified the relationship between ANT and Foucault's work, explaining that ANT can agree with Foucault's analysis of the body as site, but differs from Foucault in rejecting "deep strategies," and theorizing power in objects and networks rather than only in bodies.

Law's 2002 book *Aircraft Stories* explained in greater detail his answer to the issue of singularity: *fractionalism*. To explain how multiplicities may appear singular, Law evoked the metaphor of the fractal: something that is more than one, but less than many. He wrote that "fractional coherence" draws things together, yet does not center them (2). Part of the problem all along has been the dependence of social science on language, limiting its ability to comprehend or adequately describe fractionality. Rather than the singularity of narrative language as "fractional knowing" (4), expressed as "multiple storytelling makes rhizomatic networks...elaborations that hold together, fractionally, like a tissue of fibers" (5). In this model, intersectionality and coherence happen in the fractional areas where arrays/networks/systems participate with one another.

Working with ANT

It is impossible to separate materiality from culture and vice versa. Life is what is being done right here and now, and thus mundane and spectacular performances are equally important for describing and evoking the mattering of culture: material culture matters because we make it so. ANT provides a practical way to study this process, providing a theoretical structure that can account for the relationship of material and culture. Overcoming dualism is just one of ANT's useful contributions to nonhierarchical understandings of everyday life, seeing culture as performance and effects of performances. Much qualitative research still tends to privilege the human and thereby reify dualisms that reify notions of physicality and the metaphysical, making ANT a preferable theoretical grounding for the study of material culture. ANT looks semiotically at materials and processes of communication—translations producing immutable mobiles that endure over distance.

Language cannot escape the appearance of singularity, because in the end, if one is to use language, it is a story, a book, an object, a singular and structured translation/narrative. But singularity is an aesthetic result of the way we are able to see. To dream of escape is to write as ontological fantasy—the fantasy of being—all the while acknowledging "I am never leaving anywhere" (e.g., in de Certeau's practice of everyday life, one escapes while never leaving), since as the "networked I" I am already everywhere I can be right now. It is only I as an author who can find escape in this way, while the multiplicity of my lived experience continues to problematize the fantasy of singularity. Likewise, it is the job of the reader to work for their own escape into or from the entelechy of the text. All I can hope to accomplish in the performance of writing is to invite my readers to review my elaboration of a technological world that they might not have otherwise imagined or recognized, and thereby form an alliance with it. As Latour (1988a) put it,

> One of the tasks of sociology is to do for the masses of nonhumans that make up our modern societies what it did so well for the ordinary and despised humans that make up our society. To the people and ordinary folks should now be added the lively, fascinating, and honorable ordinary mechanism (310).

Of course, an ethnographer wishes to avoid overindulgence in the spectacular fantasy of science fiction, the fascination with malfunction and moral consequences. How things actually do work, and the moral effects, draw my attention. Latour points us toward investigating our relationships with the often mundane performances of lieutenants that do everything from organizing spare change, soaking up escaping bodily fluids, to calculating the moment our sun will die. ANT seeks to expose the intimate relations between humans and technologies, machines and machinations, actors and actants and networks. In this method, technology/artifacts/material object may be allowed to speak its own voice in as much as it may, to describe its own array of alliances, struggles, and victories.

Note

1. I assume here a reference to Liebniz's *The Monadology*, in which monads are driven by their own innate appetites to work for their own self-fulfillment, and/or perhaps the work of Alfred North Whitehead.

References

Bijker, Wiebe E. and John Law. 1992. "General Introduction." Pp. 1-23 in *Shaping Technology/Building Society: Studies in Sociotechnical Change*, edited by Wiebe E. Bijker and John Law. Cambridge, MA: MIT Press.

Carey, James W. 1989. *Communication as Culture.* London: Routledge.

Callon, Michel. 1986. "'Some Elements of a Sociology of Translation: Domestication of the Scallops and the Fishermen of St. Brieuc Bay.' Power, Action and Belief: A New Sociology of Knowledge?" *Sociological Review Monograph*, 32:196-233.

———. 1999. "Actor-Network Theory—The Market Test." Pp. 181-195 in *Actor Network Theory and After*, edited by John Law and John Hassard. Malden, MA: Blackwell.

Callon, Michel and Bruno Latour. 1981. "Unscrewing the Big Leviathan: How Actors Macrostructure Reality and How Sociologists Help Them to Do So." Pp. 277-303 in *Advances in Social Theory and Methodology: Toward an Integration of Micro- and Macro-Sociologies*, edited by Karen Knorr Cetina and Aaron Cicourel. Boston, MA: Routledge and Kegan Paul.

Callon, Michel and John Law. 1995. "Agency and the Hybrid Collectif." *South Atlantic Quarterly*, 94:481-507.

Hay, James. 2001. "Locating the Televisual." *Television and New Media*, 2:205-234.

Heidegger, Martin. 1996. *Being and Time.* Albany: State University of New York Press.

Latour, Bruno. 1986. "The Powers of Association: Power, Action and Belief: A New Sociology of Knowledge?" *Sociological Review Monograph*, 32:264-280.

———. 1988a. "Mixing Humans and Nonhumans Together: The Sociology of a Door-Closer." *Social Problems*, 35:298-310.

———. 1988b. "Part Two: Irreductions." Pp. 158-236 in *The Pasteurization of France*, edited by

Alan Sheridan and John Law. Cambridge, MA: Harvard University Press.
———. 1988c. "The Prince for Machines as Well as for Machinations." Pp. 20–43 in *Technology and Social Process*, edited by B. Elliott. Edinburgh: Edinburgh University Press.
———. 1998. "On Actor-Network Theory: A Few Clarifications." Centre for Social Theory and Technology, Keele University. Accessed December 22, 2004. Available at: http://amsterdam.nettime.org/Lists-Archives/nettime-l-9801/msg00019.html.
———. 1999. "On Recalling ANT." Pp. 15–25 in *Actor Network Theory and After*, edited by John Law and John Hassard. Malden, MA: Blackwell.
Latour, Bruno and Stephen Woolgar. 1979. *Laboratory Life: The Social Construction of Scientific Facts*. Beverly Hills, CA: Sage.
Law, John. 1986. "On the Methods of Long-Distance Control: Vessels, Navigation and the Portuguese Route to India. Power, Action and Belief: A New Sociology of Knowledge?" *Sociological Review Monograph*, 32:234–263.
———. 1992. "Notes on the Theory of the Actor-Network: Ordering, Strategy, and Heterogeneity." *Systems Practice*, 5:379–393.
———. 1999. "After ANT: Complexity, Naming and Topology." Pp. 1–14 in *Actor Network Theory and After*, edited by John Law and John Hassard. Malden, MA: Blackwell.
———. 2000. "Objects, Spaces and Others." Center for Science Studies, Lancaster University, United Kingdom. Accessed December 26, 2004. Available at: http://www.comp.lancs.ac.uk/sociology/papers/law-objects-spaces-others.pdf
———. 2002. *Aircraft Stories: Decentering the Object in Technoscience*. Durham: Duke University Press.
Lury, Celia. 1996. *Consumer Culture*. Cambridge: Polity Press.
Star, Susan Leigh. 1991. "Power, Technologies and the Phenomenology of Conventions: On Being Allergic to Onions." Pp. 26–56 in *A Sociology of Monsters? Essays on Power, Technology and Domination*, edited by John Law. London: Routledge.

3
The Social Construction of Technology (SCOT): The Old, the New, and the Nonhuman

Trevor Pinch

The Social Construction of Technology (SCOT) approach was first developed by Pinch and Bijker (1984). Its key idea is that meanings of technological artifacts are shared by social groups. Problems posed by artifacts for certain social groups are responded to by new designs. Over time the early emphasis in SCOT on explaining technology in terms of a stable society has been replaced by stressing the *mutual construction* or *mutual shaping* of technology and society. This process is evidence that technology, society, and materiality are in continuous interaction.

The recent prominence given to materiality in a number of disciplines (e.g., Appadurai 1986; Buchli 2002; Dobres and Hoffman 1994; Gumbrecht and Pfeiffer 1994; Miller 1998, 2005; Pinch and Swedberg 2008) puts on the table an important set of issues for followers of the SCOT approach and for social science analysts at large. That material objects and things deserve attention is by now obvious. It is clear that social processes of representation, enactment, identity formation, and performance are embedded within the material world of things. The importance of materiality has always been clear in the new sociology of technology, where "opening the black box" of technology has become the rallying cry (Bijker, Hughes, and Pinch 1987; Latour 1987; MacKenzie 1993; MacKenzie and Wajcman 1986; Pinch and Bijker 1984). Technologies also tend to bite back (Tenner 1996) and thus also put on the agenda issues of agency and in particular nonhuman agency (Latour 1992). A deeper consideration of certain properties of materiality is, however, necessary if SCOT wishes to refine its ideas and simultaneously to defend itself from critics.

In this chapter I address two important topics for SCOT that have arisen from the growing attention to materiality and recent work in the field of technology studies. First, historian of technology David Edgerton, in his aptly named book *The Shock of the Old: Technology and Global History since 1900* (2006), has drawn attention to the problem of older technologies coexisting with newer technologies. The coexistence of the old and the new is a particularly pressing issue for SCOT because of its roots in the sociology of scientific knowledge. If there is one defining feature of science it is that the demise of old science is coextensive with the rise of the new. But in technology the situation is different as our mundane preference is often for older technologies. Think of the modern act of taking a meal. It is replete with old technologies such as tables, chairs, eating implements, cooking utensils, and so on. Understanding the characteristics of this phenomenon is necessary for the further development of SCOT.

The other issue I take up here is a long-standing one: the role of nonhumans in the analysis of technology. Actor-Network Theory (or ANT) (Latour 1987, 1992; and see Kien this volume) brought this issue to the table by arguing for analytical symmetry between humans and nonhumans. Such work has been posed as a challenge to the traditional SCOT approach with its emphasis upon the role played by social groups of humans in the construction of technology. In a volume on material culture it is particularly important to address these issues as a lens to understanding the usefulness of SCOT as an analytical approach distinct from ANT and others.

The Old and the New

The rejection of new technologies in favor of old technologies is evidently a widespread phenomenon that needs conceptualizing within SCOT. In one sense it brings the politics of technology out in a striking and obvious way. If users favor, say, riding old bicycles over driving new cars, and brushing their teeth with a handheld tooth brush rather than a more modern electric toothbrush, they are making, in part, political choices.

In SCOT it is argued that the key aspect in understanding technological development is to identify social groups that share meanings of particular technologies. It is the *meanings* given to technics and techniques that provide a way of understanding successful, failed, tangential, and niche-market technologies. Sometimes these meanings are contested in what is called "interpretative flexibility" (a concept taken over from the sociology of science; see Collins 1992[1985]; Pinch 1986). Thus, for example, in examining the transition from the "penny farthing" bicycle (known at the time as the "Ordinary") to the safety bicycle at the turn of the nineteenth century, Pinch and Bijker (1984) identify two different meanings of the Ordinary bicycle: the "unsafe" bicycle and the "macho" bicycle. The unsafe bicycle is the meaning shared among the social groups of women and elderly men who wanted to use the bicycle for transport. The unpaved roads at the time meant that the Ordinary with its capacity to send the rider over the handle bars (known as "doing a header") was considered to be too risky for many users. Among the social group of "young men of means and verve," on the other hand, this bicycle was used for sporting purposes and for showing off (e.g., to female friends) in the parks. The transition from the "Ordinary" to the aptly named "safety bicycle" (with two similar sized wheels)—accompanied by a series of artifacts and developments (including the air tire) too complex to describe here (see Bijker 1995b)—can be understood as a change in the meaning of the bicycle as the concerns of certain social groups came to predominate. In this form of analysis engineers and the problems they address are to be thought of as responding to the needs of different social groups.

Clearly one key aspect in old technologies continuing to be used is that there must be a social group that shares a meaning of the technology *as usable*. The "Ordinary" bicycle is an interesting place to start our considerations in this chapter because the only people to ride Ordinaries today are perhaps specialized historians of the bicycle (such as Wiebe Bijker) and the only place one will find Ordinaries is in museums. If there is a social group built around the technology today, thus, it is collectors. Such a social group may share a meaning of the artifact (e.g., as an object for museum display), but such a meaning does not include the original and most important functional meaning of the use of the object. There is no current social group that shares the meaning of the Ordinary bicycle as something to ride (for either transport or sport). The safety bicycle does have such a social group, though.

For an old technic to continue to function, other things also need to be in place. As Thomas Hughes (1984, 1987) has powerfully argued, technics can be thought of as part of sociotechnical systems, or sociotechnical ensembles (Bijker 1995a). Actor-Network Theory makes a similar point by noting that technics or technological artifacts are part of sociotechnical assemblages. No technology is an island unto itself. If there were no modern paved roads from the development of the transportation system, it is doubtful that bicycle riders could use their machines for transportation at all (specialized sporting purposes might still exist, however).[1] Different transportation systems provide different amounts of infrastructural support to bicycle riders in different ways. In most towns in the United States there is hardly any infrastructural support beyond the basic paved roads. In the Netherlands, on the other hand, it is common to find special pathways for bicycles (and some motorized bicycles) with their own traffic lights, signs, and legal stipulations as to their rights. The rights aspect is important as the law often serves as the mediating point between the different functioning and capabilities of coexisting older and newer technologies. Sailing boats, which are obviously less maneuverable and speedy than the more modern motorized boats in most circumstances, have more advantage over power boats on rivers and the high seas. Even if not sanctioned by law, custom plays an important part in dealing with how the old and new coexist. Thus to this day it is customary when driving in the countryside to slow down your motor vehicle when passing a horse-backed rider. These laws, rules, and customs of course change depending on how well entrenched the new and the old technologies have become.

Thus an important set of considerations in addressing the relationship between old and new technologies is the infrastructure of the wider sociotechnical system, including other technological artifacts, material constraints, rules, regulations, and customs. Science and Technology Studies (S&TS) has increasingly paid attention to these infrastructural issues (e.g., Bowker and Star 1999; Pinch and Swedberg 2008; Star 1999). Part of a technological infrastructure may enable some of the capacities of an old technology to exist even though the material form of the technological artifact may have changed dramatically. In the case of music, recordings can be changed from medium to medium—thus enabling a sound from an earlier period to be heard in a completely different period. Thus the sound of "classic rock" associated with particular instruments and recording techniques can be heard in a new medium today. Modern synthesizers can emulate not only earlier instruments but also earlier synthesizers emulating earlier instruments.

In thinking about the macro level of infrastructure the nation-state is clearly one of the most important actors. America's decision to develop interstate highways in the 1950s paved the way (literally) for the widespread adoption of the automobile (and the disappearance of rival public forms of transportation). How different nation-states went about highway-building reveals the different politics and technical choices involved (Zeller 2007). Regulation of technologies across and between nation-states is also important. The EU regulates genetically modified food such that they must be labeled and traceable but the United States does not (Lezaun 2006), for example. Furthermore, as Edgerton shows, uneven global development and global technology flows play a crucial role. Thus an indigenous technology such as the rickshaw can be combined with a modern motor scooter to form the "creole technology" of the autorickshaw specifically in the cultural and economic context of poor Asian countries. Again such older technologies coexist with modern mass-produced motor vehicles.

Another important issue related to the materiality of infrastructure is the parts, repair, and maintenance necessary for older technological artifacts. Technics cannot continue to function unless there are replacement parts and components and social groups that are capable of repairing and maintaining them. This is a huge issue with any aging technic that still has a use. A good example is the space shuttle that now uses such aged computers that NASA regularly surfs E-Bay for old computer modules to buy. Here vagaries in technological developments between different countries can become important. For example, tubes (known as valves in the UK) are still used in the United States by enthusiasts of high-end audio and one of the few remaining sources of manufacture is Russia, where tubes were commonly favored over transistors in all sorts of technologies. Repair and maintenance have become vital issues too in keeping an operational nuclear deterrence in the United States (especially with the test ban treaty preventing testing of how nuclear weapons are currently functioning; see Sims and Henke 2008).

The tension between old and new is played out within manufacturing and marketing too. The so-called built-in-obsolescence phenomenon is something that is claimed as a necessary part of capitalist production whereby consumers are forced constantly to purchase replacement technological devices that have the same, or marginally improved, functionality. What is the natural lifespan of a product? Clearly the products of some technics (my great-grandmother's china, for instance) last longer than others' (my Cornell University souvenir pen). The decision of manufactures to no longer "support" an earlier version of a product (say, Windows 98) may in effect sound the death knell for that technic. In terms of maintenance and repair the role of users can also be crucial. Even though manufacturers may not support an old artifact, users may themselves organize to keep it alive (e.g., see Lindsay 2003). The Internet, supplemented by specialist magazines, has seen the rise of numerous use groups for the support, exchange of information, and repair of old devices ranging from old computers such as Radioshack's TSR 80 (ibid.), to long-abandoned "vintage" toys. With the rise of E-Bay, the purchase of old technics, including old Soviet-style submarines and aircraft, has probably never been easier. Interestingly this use of a new technology (the Internet) to support old technologies points exactly to the sorts of odd bedfellows and hybrid support systems one finds in this world of coexistence.

In terms of SCOT these issues of infrastructure can be thought of in terms of social groups nested around a technology that shares a meaning of that technology. Bijker's (1995b) notion of a technological frame is a parsimonious way of expressing this idea. A technological frame can be thought of as akin to Kuhn's notion of a paradigm for technology. It offers an overarching framework of shared practices, values, and meanings built around a particular technic and set of techniques that goes beyond any individual social group. Those who share, say, the technological frame of the "Broads Sloop" (a sailing boat with a flat wooden hull and giant sail [to get above the tree-line] common on the Norfolk Broads in England in the 1930s) will have a keen interest in sailing such craft (which they either own or rent), will help maintain the boats at a specialist boatyard in Norfolk, will put pressure on the Norfolk County Council School Education Board to train children to sail such boats, and regularly fundraise to maintain the boats and this traditional way of sailing. In short these enthusiasts involving many social groups, such as boat repair people and educators, share the common technological frame of the "broad sloop."

One last issue I address here is the "crosstalk" or potential interference generated by the coexistence of old and new technologies. In the case of transport technologies such as ships, cars, and bicycles special material, social, and legal remedies are used to keep them apart or to establish priority rules in particular contexts of use. A new technology such as the railway can be a direct threat to an old one, such as the canals, not only in terms of making the older technology redundant (in other words diminishing the size and number of social groups that use that technology) but also because the new one can directly damage the working of the old one. For example, organic farmers oppose genetically modified crops because of cross-pollination that threatens the genetic purity, certification, and sales of their own organic crops. If the interference problem cannot be addressed and solved, it is possible that it can lead to a demise of one or more of the technologies involved. Ultimately the interference between old and new technologies (which can occur also from overuse of the same technology within a restricted domain such as the overcrowding of the radio spectrum) needs to be recognized and dealt with by special intermediary social groups invested with legal authority. Often governmental agencies and quasi-legal regulatory bodies are set up (such as the U.S. Federal Communication Commission [FCC]; see Dunbar-Hester [2009] that deals with the radio spectrum) to legislate between the competing interests. These interference problems are likely to be hard to deal with because they may involve unintended consequences, and hence will go unrecognized by system designers and will require coordination and mediation between different social groups around a technology.

One last important way that old technics live on is when a new use is found for them. The importance of "users as agents of technological change" (Kline and Pinch 1996) and the role of users in technological innovation (Von Hippel 1988) is an important theme in current technology studies (Oudshoorn and Pinch 2003; Pinch and Trocco 2002). Sometimes a new use involves the emergence of interpretative flexibility. A good example here is the use of the turntable and vinyl LP by DJs for "scratching." Suddenly turntables (and in particular the Technics SL-1200 that is the turntable of choice) have found a new lease on life despite the demise of vinyl elsewhere. Technologies such as the turntable can in effect acquire a new meaning, to become components of a new technological system, and thereby continue to be manufactured and supported (Fouché 2006; Faulkner and Runde 2008).

The Challenge of the Nonhumans

The most controversial issue in technology studies today, and indeed a foundational issue for the ontology of technology, is how to deal analytically with nonhumans. Some approaches, such as the one advocated by Bruno Latour (1992) and Peter-Paul Verbeek (2005), argue for symmetry between humans and nonhumans. In other words, for analytical purposes, humans and nonhumans are to be treated as equivalent. The impetus for this sort of work is to move beyond the social shaping or social construction of technology by humans and to take account of the effects of technology on social formations.

I have no problem with the idea of new social groups coming into existence with the introduction of new technologies. Obviously social groups are not stable forever and new ones have to emerge from something. For example, in studying the history and social construction of the electronic music synthesizer—a whole new class of instruments that start to appear in 1964—I found that all sort of new social groups including synthesizer players, manufacturers, and salesmen came into existence (Pinch and Trocco 2002; see also Merrill this volume, in relation to recording technologies). The 1960s psychedelic movement was also partly shaped by the synthesizer, which allowed for the exploration of new washes of sound, unusual timbres, spacey effects, and so on. The recognition of the mutual interaction between social groups and technologies such as with the formation of a new social group around a new technology does not, however, necessarily lead to the more radical position that humans and nonhumans are equivalent or that every effect of technology on humans needs to be brought into the analysis. Indeed, I will argue below, approaches that call for the symmetry of humans and nonhumans miss the more important issue, which is to think about how and under what circumstances nonhumans and their impact are made visible in the first place. Let us now go into these arguments in more depth.

In calling for a new postphenomenological philosophy of technology Peter-Paul Verbeek (2005) dismisses SCOT. Simply put, he says this approach ignores the effects of technology on humans. He gives the example of how a microwave oven—a nonhuman device—has transformed eating. He writes: "the factors that determine whether human beings take their meals together include not just human beings but the microwave itself. Reducing technology to social interactions fails to do justice to the active role played by technologies themselves" (102). The effects of technologies on humans, however, are hardly uniform. As Peterson's chapter on microwave users shows, for example, many individuals negotiate the meanings and functions of microwaves or reject them altogether.

Furthermore, the phrase "reducing technology to social interactions" mischaracterizes SCOT. The SCOT approach is a way of understanding technological development that emphasizes the role played by social groups. It does not claim that there are no effects or impacts of technologies upon humans. It would have been very strange indeed if Bijker and I could have said anything at all interesting about bicycles and their history if we ignored the effects of nonhumans on humans. Clearly bicycles depend on roads (nonhumans) and being able to steer your bicycle around a bend is an example of a human responding to a nonhuman: a bend in the road. If you think of the nonhuman effects in how a bicycle works there are clearly many: the frame must be rigid enough to support riders; the handle bars must turn freely such that riders cannot only steer but react to bumps on the road; the chain must be tight enough to enable a hill to impact on riders such that they feel the need to pump their legs harder; the tires must be inflated and of the correct material to lessen the impacts of bumps in the road, and so on. Add the laws of physics and balance so that upright riding can be maintained under different conditions and you have even additional amount of nonhuman effects. In writing about the bicycle, Bijker and I were, of course, quite aware of the many ways that bicycles, roads, and so forth impact humans. We saw no need to lay stress on such effects, but our form of analysis does not deny such effects. Rather, our analysis is selective in the aspects of the nonhuman world it chooses to focus upon.

There is an enormous amount of stability in how the nonhuman world impacts humans and this is something shared across all social groups.[2] Walls impact humans but no one seriously thinks this always needs to be stated, unless you are analyzing new architectural practices or an earthquake. Likewise in the analysis of the bicycle all bicyclists whether "young men of means and verve," elderly men, or women respond to nonhumans like roads every time a bend is taken. Sometimes, however, in the process of the development of a technology, what goes on in the nonhuman world becomes much more visible as engineers and others contest the exact properties and performance of the nonhumans. Such contested meanings of the nonhumans and their attributes are exactly what SCOT tries to make visible. In the bicycle example we showed that air under pressure can be a problem for tire construction, and this directly affects the adoption of the new safety bicycle; we discussed the impact of roads upon bodies directly as in "doing a header"; we showed that bicycles with big wheels go faster and this leads to developments of extraordinarily large-wheel bicycles like the Rudge Ordinary, and so on (Pinch and Bijker 1984). In other words, we had and still have a methodology

Pinch 53

for making certain nonhumans visible. If the nonhumans are relevant to social groups, then they are relevant to the analysis.

If there are indeed many effects of a technology on humans, why settle on just one social effect of the microwave as Verbeek does—that eating patterns may change? The number of potential effects of the microwave upon humans, and the possibilities that microwaves afford, are several (see Peterson this volume). So why pick out just one effect on humans and make this the crucial one? How would we know which nonhuman effect to focus our attention upon?

Let us now examine a much more famous example in technology studies: the "sleeping policeman" (British term)—or as more commonly known, speed bumps—discussed by Latour (1992). In introducing the idea of a nonhuman policeman Latour points out that the moral force of signs is weak and that their replacement by technics, such as speed bumps, "forces" cars to slow down (for a different use and outcome, however, see Noy this volume). Latour talks about this as a process of delegating to nonhumans a form of morality formally carried out by humans. The sign "Slow Down" is replaced by a speed bump, which does the job more reliably. But all that is at stake is the point known since the dawn of eternity that nonhumans can directly impact humans—try walking into a wall, if you are not convinced. The only little twist in the speed bump example is that the interaction with the nonhuman is mediated by another nonhuman, the car, which the human will want to avoid getting damaged or risk an accident (the economic incentive to save your car from an expensive repair and safety combine here). The decision to slow down is of course made by the human and even if that decision were delegated to a machine (say a special detector on the car that notices a speed bump approaching) it will still be a human who has programmed the machine and made the decision to tell the car to slow down. People understand this all too well. The way to think about this issue is to realize that "forcing," "guiding," "steering," and so on is something being done to us all the time by nonhumans, but for the most part goes unnoticed. Every time you walk on the pavement you are being "guided" by the nonhumans that make up its physical structure. Every time you drive your car, take a bend in the road, avoid a pothole, and so on, nonhumans are enabling or constraining what you do. Normally this role played by the nonhumans is so taken for granted that it is invisible. Think about driving your car or bicycle round a bend in the road. Perhaps you have a desire or intention to get from A to B quicker by following a straight line and ignoring the bend. What keeps you on the road is, just like the speed bump does, the fear of damaging your vehicle.

The issue of delegation is one that we need to carefully address. In

Latour's analysis the intention to get cars to slow down is delegated to the speed bumps. The example is convincing because the intention is spelled out in a sign ("Slow Down") that has now been replaced by a nonhuman: the speed bump. But the problem is that such delegation is everywhere. The delegation/intentionality argument can in principle be made every time that a purposely built nonhuman artifact "forces" a human to do something. Think of all the myriad bends the road takes. We can say the road engineers constructed *that* bend *there* intentionally and that they delegated their intention to the nonhuman bank of the road, and so on. Now it may be argued that the bend in the road involves no intentionality on the part of the designers because they are just following the natural terrain of an encroaching hill. But there is truly no "natural" element in road building. We just have to be reminded of how Napoleon III rebuilt Paris in the 1850s at huge cost so as to maintain straight boulevards. Hills were tunneled through and buildings raised and lowered to ensure that a road was straight.[3] In other words a road builder who opts for a curve can be said to have intentionally avoided the straight option and delegated this intent to the bend in the road. The road itself, in that it supports the car, can be thought of as a delegation by the road designer to the road to support cars. But with this way of thinking delegation quickly becomes trivial because it is happening nearly everywhere all the time. What we need to think about is the politics of how roads actually get designed.

Conclusion: Infrastructure in Action

I will conclude with a personal example of how politics, materiality, and design interact in dealing with the old, the new, and the nonhumans. In the little hamlet where I live in the United States I have become president of my local housing association. In response to the problem of increasing vehicle traffic in our area we have been implementing a traffic-calming plan. This plan necessitates detailed discussion over the materials to be used in building road surfaces. One of the most contentious issues in our community is over something called "cobbled shoulders" that are to be used at the edges of roads, in the section of the road between the sidewalk and the main road reserved for vehicular traffic. The basic idea behind the implementation of cobbled shoulder bands is to keep cars farther away from the sidewalks than a normal road surface would. But with the change in the design of shoulders a new problem arises. The problem, as many concerned citizens ask, is "where will the cyclists go after the implementation of cobbled shoulders?"

Suddenly with all the talk of sustainability in the air the bicycle on the ground has achieved a new prominence in traffic planning. Our neighborhood is full of cyclists and many of them ride on road shoulders on their way to work as our already narrow roads and crammed-together houses have no room for Dutch or Danish style bicycle paths. After the implementation of cobbled shoulders the bicycles will have to share the road with the rest of the traffic. But how can they best do this in safety? Our traffic consultant tells us that most serious bicycle accidents are caused by bicyclists leaving bicycle lanes placed on the right of roads to turn left across the path of the traffic. It is actually safer, as the argument goes, for the bicyclists to ride in the newly calmed traffic. But how do you prevent bicyclists from riding, as usual, in the shoulder (which for them seems like the next best thing to a bicycle path)? The answer our plan calls for is to make the shoulder of a specially textured material (cobbled stones) that will give the bicyclists a bumpy "buzz" that is sufficiently uncomfortable that it will force them to ride in the main road. Bicyclists are unhappy about this, however, for this is to assume that the traffic is already sufficiently calmed that they can ride in it in safety. But what if the cars speed? As one bicyclist put it at a recent meeting, "It sounds as if you want to use bicyclists as speed bumps!" Parents also worry about children on their bicycles who will not feel confident enough to ride in traffic. The proponents of cobbled shoulders point out that the cobbled material itself is not unsafe to ride on and that in emergencies bicyclists can indeed ride on the shoulder; it will just be uncomfortable. This is not enough to calm upset bicyclists, however, who feel shortchanged by the new plan.

This is real technological politics in action. Our community is divided over cobbled shoulders. Neighbors threaten to no longer talk to each other and allegations of ill-will are in the air, while the poor president must use all his good humor and negotiating skills to bring about compromises. The problem we face is the classic one of how do older technologies, such as bicycles, fit in with newer systems endowed with greater capacities such as cars? Furthermore, rather than accept the influence of the nonhuman world upon the human, in this case the nonhuman world is deeply contested. The beauty of the cobbled shoulder example is that it makes visible (or audible) part of the vast and normally invisible world of nonhumans with which we live. In this case the human contestation makes visible what is at stake because humans do not want to accept the "forcing" character of the nonhumans. Now suppose the cobbled shoulder advocates win the debate. In years to come the cobbled shoulder will again become invisible (except to those bicyclists who curse it): another aspect of our road infrastructure. It will not be apparent that the nonhuman forcing action derives from a human decision made many years earlier. Latour, Verbeek, and others who advocate an ontological turn to technology studies need a means of sifting through all the myriad nonhumans and their countless interactions with humans so that we can see the significant choices made by humans. It is by returning to the politics of technology and the forums in which they are manifest that we are reminded of how this can be done.

When the nonhumans are contested—such as at planning meetings and the like—people engage in a form of analysis, typical of S&TS. As researchers in S&TS we are the opposite of magicians who make things and people disappear; our goal is actually to make things and the humans who interact with them reappear. The citizens of my hamlet remind us about the nonhumans we need to focus upon. It is in this carrying out of the mundane politics of infrastructure—infrastructure in action—where the old and the new and the human and the nonhuman are negotiated. Such infrastructural issues and the materiality they raise should become an important part of SCOT's future research agenda.

Notes

1. Of course specialized "off-road" mountain bikes and motor-cross bikes are designed to be used without paved roads, but even these machines depend on an infrastructure of paved roads for distribution of gas, parts, and so on.
2. One way of talking about material effects is through the language of "affordances." For an attempt to develop a social constructivist account of affordances see David and Pinch (2008).
3. I am grateful for the discussion with Park Doing and Peter Dear over this point. On Parisian road building, see David H. Pinkney, *Napoleon III and the Rebuilding of Paris* (Princeton: Princeton University Press, 1958).

References

Appadurai, A., ed. 1986. *The Social Life of Things: Commodities in Cultural Perspective.* Cambridge: Cambridge University Press.

Bijker, Wiebe. 1995a. "Sociohistorical Technical Studies." Pp. 229-256 in *Handbook of Science and Technology Studies*, edited by S. Jasanoff, G. E. Markle, J. Petersen, and T. Pinch. London: Sage.

———. 1995b. *Of Bicycles, Bakelites, and Bulbs: Towards a Theory of Sociotechnical Change.* Cambridge, MA: MIT Press.

Bijker, Wiebe, Thomas Hughes, and Trevor Pinch (Eds.). 1987. *The Social Construction of Technological Systems.* Cambridge, MA: MIT Press.

Bowker, Geoffrey and Susan Leigh Star. 1999. *Sorting Things Out: Classification and Its Consequences.* Cambridge, MA: MIT Press.

Buchli, Victor (Ed.). 2002. *The Material Culture Reader.* Oxford: Berg.

Callon, Michel and Bruno Latour. 1992. "Don't Throw the Baby Out with the Bath School! A Reply to Collins and Yearley." Pp. 343-368 in *Science as Practices and Culture*, edited by

Andy Pickering. Chicago: University of Chicago Press.
Collins, Harry. 1992 [1985]. *Changing Order* (2nd ed.). Chicago: University of Chicago Press.
David, Shay and Trevor Pinch. 2008. "Six Degrees of Reputation: The Use and Abuse of Online Review and Recommendation Systems." Pp. 341-374 in *Living in a Material World*, edited by Trevor Pinch and Richard Swedberg. Cambridge: MIT Press.
Dobres, Marcia-Anne and Christopher Hoffman. 1994. "Social Agency and the Dynamics of Prehistoric Technology." *Journal of Archaeological Method and Theory,* 1:211-258.
Dunbar-Hester, Christina. 2009. "'Free the Spectrum!' Activist Encounters with Old and New Media Technology." *New Media & Society,* 11 (forthcoming).
Edgerton, David. 2006. *The Shock of the Old: Technology and Global History since 1900.* London: Profile Books.
Faulkner, Phil and Jochen Runde. 2008. "On the Identity of Technological Objects and Innovations in Function." *Academy of Management Review* (forthcoming).
Fouché, Rayvon. 2006. "Say It Loud, I'm Black and I'm Proud: African Americans, American Culture, and Black Vernacular Technological Creativity." *American Quarterly,* 58: 639-661.
Gumbrecht, H., and K. Pfeiffer (Eds). 1994. *Materialities of Communication.* Palo Alto: Stanford University Press.
Hughes, Thomas Park. 1984. *Networks of Power: Electrification in Western Society, 1880-1930.* Baltimore: Johns Hopkins University Press.
——. 1987. "The Evolution of Large Technological Systems." Pp. 51-82 in *The Social Construction of Technological Systems: New Directions in the Sociology and History of Technology,* edited by Wiebe Bijker, Thomas P. Hughes, and Trevor Pinch. Cambridge, MA: MIT Press.
Kline, Ronald and Trevor Pinch. 1996. "Users as Agents of Technological Change: The Social Construction of the Automobile in the Rural United States." *Technology and Culture,* 37:763-795.
Latour, Bruno. 1987. *Science in Action.* Milton Keynes, UK: Open University Press.
——. 1992. "Where Are the Missing Masses? The Sociology of a Few Mundane Artifacts." Pp. 225-258 in *Shaping Technology/Building Society,* edited by Wiebe Bijker and John Law. Cambridge, MA: MIT Press.
Lezaun, Javier. 2006. "Creating a New Object of Government: Making Genetically Modified Foods Traceable." *Social Studies of Science,* 36:499-531.
Lindsay, Christina. 2003. "From the Shadows: Users as Designers, Producers, Marketers, Distributors, and Technical Support." Pp. 29-50 in *How Users* Matter, edited by Nelly Oudshoorn and Trevor Pinch. Cambridge MA: MIT Press.
MacKenzie, Donald. 1993. *Inventing Accuracy.* Cambridge, MA: MIT Press.
MacKenzie, Donald and Judy Wajcman. eds. 1986. *The Social Shaping of Technology.* Milton Keynes: Open University Press.
Miller, Daniel (Ed.). 1998 *Material Cultures: Why Some Things Matter.* Chicago: University of Chicago Press.
——. (Ed.). 2005. *Materiality.* Durham: Duke University Press.
Oudshoorn, Nelly and Trevor Pinch. 2003. *How Users Matter.* Cambridge, MA: MIT Press.
Pinch, Trevor. 1986. *Confronting Nature: The Sociology of Solar-Neutrino Detection.* Dordrecht, The Netherlands: Kluwer.
Pinch, Trevor and Wiebe Bijker. 1984. "The Social Construction of Facts and Artifacts." *Social Studies of Science,* 14:399-441.
Pinch, Trevor and Richard Swedberg. 2008. *Living in a Material World: Economic Sociology*

Meets Science and Technology Studies. Cambridge, MA: MIT Press.
Pinch, Trevor and Frank Trocco. 2002. *Analog Days: The Invention and Impact of the Moog Synthesizer*. Cambridge, MA: Harvard University Press.
Sims, Benjamin and Christopher Henke. 2008. "Maintenance and Transfer in the US Nuclear Weapons Complex." *IEEE Technology and Society Magazine*, 27:32-38.
Star, Susan Leigh. 1999. "The Ethnography of Infrastructure." *American Behavioral Scientist*, 43:377-391.
Tenner, Edward. 1996. *Why Things Bite Back: Technology and the Revenge of Unintended Consequences*. New York: Knopf.
Verbeek, Peter-Paul. 2005. *What Things Do: Philosophical Reflections on Technology, Agency and Design*. University Park: Pennsylvania State University Press.
Von Hippel, Eric. 1998. *The Sources of Innovation*. Oxford: Oxford University Press.
Zeller, Thomas. 2007. *Driving Germany: The Landscape of the German Autobahn, 1930-1970*. Oxford: Berghahn Books.

4
Material Culture and Narrative: Fusing Myth, Materiality, and Meaning

Ian Woodward

The term "material culture," now a more or less coherent field of interdisciplinary study, refers to the fact that material things are one part of culture and that they do cultural work. Being good to think with, objects are cultural categories materialized. On the importance of objects and materiality for culture there is wide agreement among social theorists, even ones who we would not necessarily see as principally studying material culture such as Marx and Durkheim. What both Marx and Durkheim understood in unique ways was that we cannot know ourselves, nor can we be participants in our own culture, without "bringing in" objects and material things. Prefiguring the current round of materiality studies, for example, Marx (1975:329) commented that it is in the fashioning of the material world that human beings come to know themselves. He saw that people constitute themselves intellectually and physically through the material things that they create and encounter.

On the other hand, Emile Durkheim (1995), the French sociologist of the sacred and profane aspects of everyday life, argued that to express our ideas and to understand ourselves we need to attach those ideas to material things that symbolize them (228-229). For any powerful, meaningful object, there needs to be a fusion between thing and idea (or words), whereby the object incarnates some type of social ideal. Durkheim emphasized that material things provide the physical means for assigning and confirming our values and attitudes. For example, for some people the I-pod is a symbol of mobile, aural-aesthetic transcendent pleasure, a precious totem in urban settings; the team football jersey is a sacred object of belonging and commitment to one's tribe and to the tribe's history, both victorious and tragic; and the little black dress is a materialization of elegance and European style. In sum, in things external to us, we have an opportunity to "objectify" (see Vannini chapter one) ourselves and our ideals.

This chapter addresses the broad question of what gives power to material things by way of narrativization. To narrate is an essential aspect of being human. Everyone narrates; narratives are basic structures of everyday perception and action. We use narratives to understand what happens in our own lives, the lives of others, and the world at large. Narratives constitute stories. Narratives are not necessarily untruths, but they follow the logic of stories or trope. Narratives allow us to piece together events, making sense of disparate experiences as if they were part of an understandable whole. Importantly, narratives frequently come to life by being embodied in objects. What is even more interesting is that objects frequently structure the very way narratives unfold. Objects stimulate narratives, or they afford us access into them. For example, a photograph reminds us of our past; the who, what, where, and when of our personal stories; a public memorial lets us understand the elemental myths, stories, and something of the folklore of a collective of people. In sum, objects acquire cultural meaning and power in the context of stories or narratives that locate, value, and render them visible and important. Without such narrative storylines—be they accounts spoken by individuals or accounts that hold more general sway within a population such as a discourse—an object is rendered virtually invisible within a culture.

This chapter is focused on the narrativization of objects through a couple of main processes. Initially, there is the way people talk about objects as a way of talking about their lives, values, and experiences. The way objects acquire their cultural meaning is partly a matter of their material form and its affordances, but an object's meaning is also constructed within local settings, where participants confer objects a social life through offering active, creative accounts, or narratives. Likewise, an object's power is also partly acquired through the culture-wide stories or myths it incarnates. Thus, an important starting principle is that it is stories and narratives that hold an object together, giving it cultural meaning. Rom Harré (2002) proposes a number of illuminating principles for theorizing objects. Two of these are (1) an object is transformed from a piece of stuff definable independently of any story line into a social object by its embedment in a narrative; and (2) material things have magic powers only in the contexts of the narratives in which they are embedded (25). This leads to the conclusion that there is a relational quality to the entanglement of social action with things and words, something summed up neatly by Pels, Hetherington, and Vandenberghe (2002:11) who write that "objects need symbolic framings, storylines and human spokespersons in order to acquire social lives; social relationships and practices in turn need to be materially grounded in order to gain temporal and spatial endurance."

In what follows I outline three aspects of the entwinement of narratives and materiality. In doing so, I delineate a schema for understanding the types of narratives that embed objects and in which objects are embedded. Though it is a schematic discussion, readers must realize that in the end it is never so simple to make such delineations because there is frequently a polysemic, dialogic process of communication that often involves all three aspects of narrativization, each playing a role depending on local conditions. The three aspects of narrativization are:

1. The narration of objects by individuals or groups; referring to the way a person might talk about objects, their meanings, and their history, incorporating them into stories that signify aspects of individual or group identity, for example, within personal, familial, or subcultural lifeworlds.
2. The location of objects in larger cultural narratives; referring to the way objects become part of the deep generative narratives of culture. For example, the electric chair helps us to narrate aspects of fair punishment, deviance, and the limits of the state; the flag helps us identify the mythological aspects of belonging to national collectives.
3. Objects that narrate; referring to the capacity of certain objects to narrate human activity directly, for example, scientific objects like weather instruments, computer screens, or GPS navigational systems that tell their own stories of particular aspects of natural and human activity.

Focusing on the first dimension of this schema, the chapter considers the various ways in which narratives give objects social life. Using the home and personal space as an exemplary research site, the chapter shows how a narrative approach can be applied to understand objects, visuality, and the capacity of words and things to make representations within personal spaces such as the home. Finally, it should be made clear that narrative approaches to material culture—while borrowing extensively from the corpus of narrative theory proper—are easily combined, as is the case with any of the theoretical approaches discussed in part one of this volume, with other analytical perspectives that emphasize culture as process.

Narrating Objects

Objects frequently require accounting for, or narrativization. As I have pointed out, narratives refer generally to stories or accounts that are told both individually and at the macrosocial, or collective, level (e.g., collective memory, history, etc.). At an individual level, narratives consist of the accounts or stories people tell themselves and others in order to both make sense of—and make through practical means—their lives. Narratives are thus reflective, accounting for events that have already taken place. But they are also active as a site for articulating an individual's values and beliefs, as they provide the schematic resources and frames for constructing a person's future. Individuals tell their lives through stories, though these stories are not simply there waiting to be told—they are actively constructed for particular audiences, plots, and contexts (Riessman 1993). This process of narrativization tells us a great deal about the meanings people apply to their lives. Narratives are told by oneself about one's relationships with others, whether they be objects, institutions, or other people (ibid.; Young 1989). Narratives are formed in private psychological space (Shotter 1984), but revealed through talk, a public process par excellence. They abide by particular rules or tropes, are internally consistent, and serve to make things socially and culturally warrantable (Riessman 1993; Shotter 1984). In Garfinkel's terms (1967:vii), they comprise "accounts" that assist in making events, objects, and behaviors rational, reportable and, above all, meaningful.

In settings and spaces where the meaning of objects is open to interpretation or debate, or where people are anxious about "fixing" the meaning of objects, there is a strong need to narrate the object. One example of where this will happen is in the field of consumption behaviors, especially in relation to forms of consumption that are expressive of status and identity, such as areas of fashion, home decoration, or knowledge of wine. In these fields choices and interpretations proliferate whereas formal knowledge is relatively scarce, aesthetic rules governing these fields are hard to claim expertise in, and consumers rely on whatever narratives they can (often these relate to the advertising narratives at hand) to make sense of objects (Warde 1994). For example, we may know little about styles of wine and the ways of describing the taste of wine, but may feel compelled in company to comment on or talk about the choice of wine at hand. Or, having guests enter our home, we may feel the need to talk about some aspect of our home—the color of the walls, the pieces of furniture at hand, the paintings on the wall. Knowing or being able to fix the meaning of these things is important because we agree that as objects they can come to represent us and, in part, we are judged by them. In my (Woodward 2001, 2004, 2007) studies of people's narratives about their home, the issue of anxiety arises in the context of research interviews with householders. Objects like ovens, refrigerators, or sculptures displayed in the home require narration because the owner is anxious to fix the meaning of the object in the particular social setting. Thus, in order to assuage possible negative judgments about one's large and expensive oven, a female participant admits that her new oven might represent materialistic or antifeminist values, but in reality it should be seen as a focus for encouraging family togetherness and as a device for the skillful preparation of artistic meals. In this instance, concerns about social status and perceptions of identity necessitate storytelling about one's choices, resulting in the narrative fixing of the meaning of objects.

Hurdley (2006, 2007) also focuses on narrative in the home, but in her interviews she investigates mantelpieces and the objects that they hold as meaningful. Mantelpieces are often a central display feature of many homes and are an architectural convention particularly suited to presenting highly personal and meaningful objects. Participants in her study showed self-awareness of how they could employ various narratives to interpret or explain an object on their mantelpiece, depending on circumstances like the degree of familiarity of visitors (2006:721). Hurdley's research illustrates the way people narrate domestic objects in the context of social roles, narrative figures (e.g., "the good grandmother"), and matters of social performance and display. Objects within the home offer an opportunity for people to tell stories about themselves, to fuse aesthetics with morality, and to deal with both immediate social contexts and historical events. Money's (2008) study of material culture in the living room delineates three key ways in which people narrate objects in their living rooms: as matters of familial obligation, as markers of memory, and as commemorative objects. Money shows how the display of objects in one's living room is often not under an individual's perfect control, as displays are used in order to accomplish a range of different social obligations and purposes, such as displaying gifts from others in recognition of the relationships they symbolize.

The role of objects within the home as commemorative and memorials is important. Even when an object is not displayed formally on a mantelpiece, its presence acts to remind residents of their connection to people, places, and events in the past. In this sense, objects can serve as a trigger for narrativization. Marcoux (2001) investigates the ritual of *casser maison* or emptying the home, which occurs when elderly people leave their homes to live in residential care settings. A similar ritual has to occur after a death, with relatives taking on the role of deciding the distribution of objects to family, friends, or disposal of them as rubbish. Marcoux's research shows that when a home has to be broken up, part of the grieving process involves distributing the cherished content of the home to family members and close friends. In this way, such objects come to stand for the presence of the absent person. Objects given away thus involve the distribution of a bit of oneself and lead the reconstruction of self in another's home. In this way, homes become accretions of memory, of social relationships with others, and of the material references to the bonds, connections, and ties one has had with others. Gibson's (2008) research into objects, death, and grieving also points to this conclusion. On the basis of her research she argues that objects that are handed down can transpose themselves into quasi-subjects, acting as a material replacement or symbol of the person who has died. In the above examples, users and possessors of objects take an active voice in narration, telling the history, associations, and trajectories of an object and so try to, at least temporarily, discursively fix an object's meaning. In the next section, moving from the personal to the collective, I discuss how culture can provide narrative frames for interpreting objects and how these narratives inform our perceptions and relations with objects.

Objects and Cultural Narratives

The second way in which narrative is relevant to our concerns is they are not only mentalistic or idealistic aspects of selfhood, but are also important components of culture. Narratives are not just told by individuals to others or to oneself. Narratives also circulate within culture, telling members of a group about their own culture, and thus also informing the use and interpretation of particular objects. Recent work within the field of cultural sociology has been important in emphasizing the textual and narrated quality of culture (see Alexander 2003). This body of work has emphasized the way the "big stories" of everyday life (e.g., in relation to values of good and bad, sacred and profane, democratic and undemocratic, beautiful and ugly) actually constitute the discursive contours of social life and how we use these discourses and work within them to create meaningful worlds.

Within the field of semiotics the work of Barthes is one important starting point for this aspect of object narration. As Vannini introduces in chapter one, Barthes's idea is that people were exposed to collective myths through their engagement with a multitude of everyday objects and experiences, from motor vehicles, to white goods, implements like cricket or baseball bats, to shoes and clothing. Myth was constituted, represented, and experienced through engagement with everyday objects. What exactly, in Barthes's terms, was a myth? Barthes (1993:109) clarifies that myth is a system of communication, or message. It is thus more than merely a physical object, but is a mode of signification (constituted by narrative, context, and story line) that is attached to objects. Myth thus adds a certain narrative patina to objects, endowing them with special qualities and abilities. Consider mythical objects like the guitar of Hendrix or the trumpet of Miles Davis, the boxing gloves of Ali, the iconic white jumpsuit of the late Presley, or a dress once worn by Marilyn Monroe. Myth thus equals materiality plus context, story line, history; or, narrative. Barthes's famous series of essays in *Mythologies* provides plenty of examples of how cultural narratives are embedded within objects and how objects come to perform or incarnate cultural myths: the new Citroen motor vehicle represents bourgeois desire, the spirit of modernity, and a beauty apparently constructed beyond human hands; the Eiffel Tower represents the attractions and myths of Paris, is a universal symbol of mobility and travel, and a materialization of the aspirations of the modern city; red wine symbolizes masculinity, the earth, French national identity, and so forth.

Given the current interest in materiality in the context of consumption practice, there are an increasing number of studies that utilize a Barthesian approach to the narrative power of myth, but adopt a less formal, more updated form of critical semiotics. The collection of essays on the Hummer motor vehicle (Cardenas and Gorman 2007), for instance, locates the behemoth automobile within the broader myths and narratives of life in the contemporary United States. Asking "why here?" and "why now?" the authors assert that the material form of the Hummer (a hypermasculine, dominating, ultraassertive, individualistic mobile fortress) seems to fuse perfectly with dominant contemporary anxieties and discourses of American, and Western, culture. These include the privatization and fortressization of public space and an associated fear of street crime, the fear of loss of American hegemony in a post-Cold War world, a desire to avoid or overcome any potential catastrophes such as climatic extremes, terror, or war and a desire to maintain a form of rugged individualism in a world melting into a proliferation of minorities and subcultures. Thus, the vehicle itself constitutes an understandably popular response to particular, culturally constituted concerns. It comes to materially embody a (seemingly) viable response to the narratives circulating within a globally shaped, national culture.

A few decades earlier the mythological aspects of the Volkswagen Beetle were also constructed by marketers (see Holt 2004), who drew not just on matters of utility and vehicle performance, but powerful cultural myths that narrated the Beetle. These campaigns championed both the iconoclastic and individualist aspects of driving the Beetle. Narratives of the Beetle managed to assuage the consciences of those who valued collectivism, modesty, and environmental care while also appealing to people's desire to be seen as creative and self-fashioning individuals. In similar fashion, Belk and Tumbat's (2005) study reveals the narratives that underpin the "cult" of the Apple brand, providing a series of story lines that narrate key features, perceived and real, of Apple products. Again, their study shows that the success of the Apple brand is due in large part to the successful embedment of the object within broader cultural narratives, connecting Apple goods to powerful cultural stories around heroes, saviors, and creationism, which are played out in the context of various pieces of Apple technology. The materiality of Apple products is important, but it is in the mythologies surrounding the brand where the most potent cultural power resides.

Moving away from arenas of consumption and looking to the examples of punishment, Smith (2003, 2008) focuses on the embedment of technologies of punishment (e.g., the electric chair, the guillotine) within cultural narratives. Far from being dry, rational objects that simply mete out punishments, Smith shows how these objects are immersed in broader cultural narratives often based around elements of culture we would often prefer to repress. His historical study shows precisely how the guillotine was understood—or "framed"—through particular cultural narratives about science, medicine, the human body, and spirit. Similar to Foucault, Smith makes the point that punishment technologies have moved in unison with the general tendencies of modernization: toward rationality, reason, and efficiency. However, Smith takes issue with the Foucauldian thesis by looking at the guillotine case within the field of punishment, and assembles historical evidence from eighteenth-century newspapers, pamphlets, and encyclopedias to show that a counterdiscourse accompanies this "rationality" discourse that is highly emotional, profoundly symbolic, and founded upon grotesque and Gothic imagery.

Smith shows that there were divergent cultural discourses surrounding the guillotine. He does not reject Foucault's emphasis on the guillotine as representing a rational instrument of science, but argues that such a thesis has displaced, even obliterated, a more culturally sensitive account of punishment. Smith argues that his historical material shows that this object was a magnet for a range of cultural discourses, some rational and functional, some irrational and emotion-charged. The objects thus offered dual narratives: a positive story of rational, efficient punishment, and a gothic story about the horror and fear accompanying the guillotine's mode of death:

> The guillotine was a "scientific" instrument; [in] that its operations involved a routinisation of bodily activity; and [in] that an emergent professional gaze relentlessly interrogated its embodied effects. Yet it will also become apparent that none of these came at the expense of more profound symbolic resonance. Far from eliminating or replacing vital symbolisms, the guillotine and its bodies became, in Levi-Strauss's terms, bonnes á penser [good to think with] for a new set of mythologies (Smith 2003:30).

Consider a final example of how objects are situated within broader cultural narratives. Alexander's (2003) study of the computer shows the cultural narratives that existed over time about this new object, and how it was narrated through public discourse alternately as a savior, or a threat. Alexander begins by laying out a key myth of modern life: the belief that science and rationality make the world a problem of purely technological means. Put another way, there is (or was, especially through most of the twentieth century) a cultural narrative that articulates the idea that technological advancements will allow us to face and overcome any problems we encounter related to time, space, and the natural elements. The modern world is inextricably directed by technologies that routinely allow us to accomplish tasks and goals human bodies alone are unable to achieve. But the story of technology that is functional, utilitarian, and a savior is not the only story to be told about technology: "Technology is rooted in the deepest resources and abysses of our imagination. It is religion and anti-religion, our god and our devil, the sublime and the accursed" (179). Alexander asserts that understandings of technology have generally been identified as a materialist thing par excellence, "the most routine of the routine" (180), that helps us to make a way and do business in the world. Yet, Alexander asserts that technology must be situated in a cultural order: a material thing, "it is a sign, both a signifier and signified, in relation to which actors cannot entirely separate their subjective states of mind" (180).

In terms of the specific meanings attributed to technology, Alexander uses the Durkheimian idea that humans divide the world into things and events that are sacred and profane. The sacred refers to images of the good that people strive to protect, while the profane describes images of evil from which people need protection. He illustrates that at the heart of our imaginings of the computer are a set of deeply felt oppositions, whereby machines have a capacity to embody both the hopes and fears generated by industrial society. Using a textual analysis of newspaper and magazine material from around the time of the computer's birth, Alexander shows how it is understood in both sacred and profane terms. As a sacred object it is a superbrain, superhuman, the closest thing to God, and something that can solve in a flash things that have baffled generations. As a profane object, the computer is a colossal gadget, a Frankenstein monster, a mathematical dreadnought, and a figure factory. Our understanding of the computer is thus informed by narratives that construct it as having the capacity to be both savior and destroyer: as a sacred thing it "is the vehicle for salvation" while its "profane side threatens destruction" (191).

The category of monuments and memorials are also an important example of how objects reference particular historical experiences and link them to social ideals and collective desires. In this sense, monuments allow us entry into the narratives of a collective. In a paper that investigates the career or biography of Auguste Rodin's sculpture "The Burghers of Calais," Richard Swedberg (2005) illustrates the trajectory of a strongly modern, utopian public artwork. Originally commissioned in 1884 by the City of Calais and first shown in 1895, the sculpture celebrates a local heroic act. The interesting question Swedberg deals with is why has Rodin's sculpture, based on a local story, been used as a monument throughout the world, including many nations in Europe, North America, Asia, and Australia? First, Swedberg argues that the monument gained popularity in its local context because it helped to forge an identity for the city of Calais, and also functioned as a representation of a collective ideal of fraternity and solidarity for the nation of France, embodying a way to understand and define painful histories and glorious futures. More generally, the sculpture appeals to honorable sentiments within modern individuals, related to a form of civil courage that helps to forge collective solidarity. The sculpture reminds ordinary citizens that to cultivate and maintain a civil society involves ordinary people making a variety of sacrifices for the good of others. Such sacrifices may be of the most mundane sort, but they may also be linked to heroism of ordinary citizens: the sculpture tells a very modern tale of social solidarity, suggesting to citizenry that "the heroes of our time are not like the lonely and extraordinary individuals that you find on the top of pedestals in sculptures from earlier centuries" (64). This explains how the sculpture became a highly mobile—perhaps iconic—representation, tapping into a narrative that appeals cross-nationally to the heart of the modern spirit. In the next section, we move from culture to materiality.

Objects That Narrate

Finally we must consider, at least briefly, the possibility that objects themselves can actively narrate. While this suggestion involves a number of dilemmas, which I will outline shortly, it is indeed possible that certain categories of technological objects actively may narrate human experience. Knorr Cetina (1997:1) outlines the "objectualization thesis," the strong version of which implies that "objects displace human beings as relationship partners and embedding environments, or that they increasingly mediate human relationships, making the latter dependent on the former." We can see how this development might give an object the power to narrate, to actively tell the story a human would normally tell. For example, the satellite navigation systems in a car offer a form of narration to the driver and passengers. Such technologies offer visual, acoustic, and numeric guidance on route, location, and direction. Increasingly, users are demanding increased narrative qualities to their navigation systems, evidenced by the popular capacity to download celebrity voices that add character and humor to driver instructions. This example raises questions about the degree to which the satellite navigation system genuinely offers narration, or whether it merely presents constantly unfolding data that, although useful reporting for incorporating into a narration, do not constitute a narration in their own right. For example, if we were to call this narration, we would have to admit that it would be a very simple and mundane type. If narration is indeed defined as storytelling, a satellite navigation system tells a very elementary and relatively uninteresting story, unlikely to keep a person interested even when celebrity voices such as

Michael Caine or Elvis Presley provide the instructions.

Cetina and Bruegger's (2000) analysis of postsocial relations in financial markets illustrates how certain financial markets "exist" on a computer screen and how the screen narrates the market. They argue for the loosening up of conceptions of sociality and for the role of objects—in this case computer-generated aggregates of anonymous human behaviors that constitute certain financial markets—in holding human relationships and embedding the self. The computer screen thus becomes the focus on human attention, and members of the trading community treat the screen as their entry point into a stand-alone world where the market is "seen" (Mayall 2006). What we have here may not constitute a traditional form of object-oriented narrative as we have seen in relation to household objects or larger cultural narratives. However, new approaches and new theoretical frameworks demand that we move our attention to the way narrative can be represented in new forms, such as the computer screen or the iconic map, such as in Vertesi's (2008) discussion of London's "Tube Map." Here, it is in the interface of the graphic representation of the city and the user's interpretation of the visual language and its reconciliation with the streets of London that we can see space for a new language of visual-technical narrative.

Conclusion: The Case of the Home

The home is a classic site where we can find a mixture of relationships between objects and narratives. Indeed, sometimes objects in the home narrate (the computer screen on the front of the fridge that monitors home microclimates or security), sometimes they require narration (such as those on a mantelpiece), and of course the idea of home exists within broader cultural narratives associated with family, gender roles, and partnerships. But, in some situations, these narrative relationships are dialogically ordered; for example, objects on a mantelpiece can narrate as well as be narrated. They may require human storytelling, but equally they may tell stories on their own. The home is therefore an interesting and important special case to conclude the discussion in this chapter of a tripartite distinction about objects and their narratives.

When considering the internal space of a home, it can be observed that domestic spaces are not exclusively public or private, as such meanings shift and slide according to the social and familial relationships of visitors to the spatial organization of the home (Lawrence 1987). This apparent fluidity of the spatial expression of the public-private distinction may necessitate shifts in narrative authorship. Objects within the home, too, serve shifting purposes according to the needs of the situation. Objects sometimes have a public role in the home as a signifier of status, style, or taste. At other times they serve as a focus for managing and narrating self-identity, family relations, or self-esteem (Csikszentmihalyi and Rochberg-Halton 1981).

Environmental psychologists have typically emphasized the psychic dimensions of home. Not only are domestic interior spaces to be played with via a person's environmental preferences (Bonnes et al. 1987), they are spaces of familial and friendship-based interaction. At a deeper level, homes are seen as "warehouses of personal experience" (Lawrence 1985:129). As Bachelard (1994) points out, the home is a site that represents a basic and important division in geographical space between the house (self) and nonhouse (nonself or other). In the environmental psychology paradigm, homes are sites for the application of resources directed toward the maintenance of self-identity and self-esteem, family relations, and notions of insiders/outsiders (Laumann and House 1970; Lawrence 1985, 1987; Rapoport 1969). But in addition to their psychological dimension, homes carry a freight of sociological meaning. Ways of living in the home, and the organization and selection of the system of objects within its spaces, are circumscribed by moral prescriptions associated with family, gender, and class positions (Madigan and Munro 1996).

This takes us back to the key argument of this chapter. Having outlined three ways in which objects and narratives are entwined, part of the key to understanding materiality is the realization that objects, though undeniably material, are always enmeshed in a variety of private and larger cultural narratives. Important personal and cultural objects transcend their material form. They have an extramaterial quality, the basis of which is provided by mythological narratives that confer objects as sacred within the context of particular social spaces and consecrate them via practices. Without the fusion of myth and meaning with materiality, objects are just assemblages of materials.

References

Alexander, Jeffrey. 2003. *The Meanings of Social Life. A Cultural Sociology*. Oxford: Oxford University Press.
Bachelard, Gaston. 1994. *The Poetics of Space*. Boston, MA: Beacon Press.
Barthes, Roland. 1993. *Mythologies*. London: Vintage.
Belk, Russell and Gulnur Tumbat. 2005. "The Cult of Macintosh." *Consumption, Markets and Culture*, 8:205–217.
Bonnes, M., M.V. Giuliana, F. Amoni, and Y. Bernard. 1987. "Cross-cultural Rules for the Optimization of the Living Room." *Environment and Behaviour*, 19:204–27.
Cardenas, Elaine and Ellen Gorman (Eds.). 2007. *The Hummer. Myths and Consumer Culture*. New York: Lexington Books.
Csikszentmihalyi, Mihaly and Eugene Rochberg-Halton. 1981. *The Meaning of Things:*

Domestic Symbols and the Self. New York: Cambridge University Press.
Durkheim, Emile. 1995. *The Elementary Forms of the Religious Life.* London: Allen and Unwin.
Garfinkel, Harold. 1967. *Studies in Ethnomethodology.* Englewood Cliffs, NJ: Prentice-Hall.
Gibson, Margaret. 2008. *Objects of the Dead.* Melbourne: Melbourne University Press.
Harré, Rom. 2002. "Material Objects in Social Worlds." *Theory, Culture and Society,* 19(5/6):23–33.
Holt, Douglas. 2004. *How Brands Become Icons: The Principles of Cultural Branding.* Boston, MA: Harvard Business School Press.
Hurdley, Rachel. 2006. "Dismantling Mantelpieces: Narrating Identities and Materializing Culture in the Home." *Sociology,* 40:717–733.
———. 2007. "Focal Points: Framing Material Culture and Visual Data." *Qualitative Research,* 7(3):355–374.
Knorr Cetina, Karen. 1997. "Sociality with Objects: Social Relations in Postsocial Knowledge Societies." *Theory, Culture and Society,* 14:1–30.
Knorr Cetina, Karen and Urs Bruegger. 2000. "The Market as an Object of Attachment: Exploring Postsocial Relations in Financial Markets." *Canadian Journal of Sociology,* 25:141–168.
Laumann, Edward and James House. 1970. "Living Room Styles and Social Attributes: The Patterning of Material Artifacts in a Modern Urban Community." Pp. 189–203 in *The Logic of Social Hierarchies,* edited by Edward Laumann, P.M. Siegel and R.W. Hodge. Chicago, IL: Markham.
Lawrence, Roderick. 1985. "A More Humane History of Homes: Research Method and Application." Pp. 113–132 in *Home Environments,* edited by I. Altman and C.M. Werner. New York: Plenum Press.
———. 1987. "What Makes a Home a House?" *Environment and Behaviour,* 19:154–68.
Madigan, Ruth and Moira Munro. 1996. "House Beautiful: Style and Consumption in the Home." *Sociology,* 30:41–57.
Marcoux, Jean-Sebastien. 2001. "The Casser Maison Ritual." *Journal of Material Culture,* 6:213–233.
Marx, Karl. 1975. *Economic and Philosophical Manuscripts of 1844.* New York: International.
Mayall, Margery. 2006. "'Seeing the Market': Technical Analysis in Trading Styles." *Journal for the Theory of Social Behaviour,* 36:119–140.
Money, Annemarie. 2008. "Material Culture and the Living Room. The Appropriation and Use of Goods in Everyday Life." *Journal of Consumer Culture,* 7:355–377.
Pels, Dick, Kevin Hetherington, and Frederic Vandenberghe. 2002. "The Status of the Object. Performances, Mediations, and Techniques." *Theory, Culture and Society,* 19(5/6):1–21.
Rapoport, Amos. 1969. *House Form and Culture.* Englewood Cliffs, NJ: Prentice-Hall.
Riessman, Caroline. 1993. *Narrative Analysis.* Thousand Oaks, CA: Sage.
Shotter, John. 1984. *Social Accountability and Selfhood.* New York: Basil Blackwell.
Smith, Philip. 2003. "Narrating the Guillotine: Punishment Technology as Myth and Symbol." *Theory, Culture & Society,* 20:27–51.
———. 2008. *Punishment and Culture.* Chicago: University of Chicago Press.
Swedberg, Richard. 2005. "Auguste Rodin's *The Burghers of Calais*: The Career of a Sculpture and Its Appeal to Civic Heroism." *Theory, Culture and Society,* 22:45–67.
Vertesi, Janet. 2008. "Mind the Gap: The London Underground Map and Users' Representations of Urban Space." *Social Studies of Science,* 38:7–33.
Warde, Alan. 1994. "Consumption, Identity-Formation and Uncertainty." *Sociology,* 28:877–

898.

Woodward, Ian. 2001. "Domestic Objects and the Taste Epiphany. A Resource for Consumption Methodology." *Journal of Material Culture*, 6:115–136.

———. 2004. "Divergent Narratives in the Imagining of the Home amongst Middle-Class Consumers: Aesthetics, Comfort and the Symbolic Boundaries of Self and Home." *Journal of Sociology*, 39:391–412.

———. 2007. "Investigating the Consumption Anxiety Thesis. Aesthetic Choice, Narrativisation and Social Performance." *The Sociological Review*, 54:263–282.

Young, Katherine. 1989. "Narrative Embodiments: Enclaves of the Self in the Realm of Medicine." Pp. 152–165 in *Texts of Identity*, edited by John Shotter and Kenneth Gergen. London: Sage.

5
Material Culture and Technoculture as Interaction

Phillip Vannini

Culture is what people think and *do* together (cf. Becker 1986; Ingold 2000). Such a focus on collective doing and making and the materiality and consequentiality of action-based cultural processes is what sensitizing concepts such as "material culture" and "technoculture" are meant to highlight. Conceptualizing culture as action and interaction is intended to downplay the importance of cognitive cultural dimensions such as values, beliefs, codes, and ideas and to emphasize instead bodily engagements, techniques, skills, habits, and the materiality of the world of interaction. The scope of this chapter is to survey the ontological foundations of such ideas and therefore of perspectives that view material culture and technoculture as interaction. By taking some license in blurring boundaries among theoretical traditions, in what follows I review four basic principles of pragmatism, symbolic interactionism, performance theory, and social semiotics.

The chapter is divided into four parts. Each part reviews one of the four principles that distinguish this pantheoretical perspective: ecology, diffused agency, emergence, and semiotic power. While to some extent these four principles are inspired by David Snow's (2001) well-known articulation of the four premises of symbolic interactionism, by modifying them, as I do, I wish to engage in a twofold extension. The first extension is pantheoretical: I find these four principles to characterize perspectives other than symbolic interactionism as Snow does, and thus I expand the range of theoretical traditions identifiable as interactionist—whether the authors of the ideas I survey explicitly invoke their interactionist roots or not. The second extension is substantive: whereas Snow's principles and remarks are primarily sociological, I attempt to develop an interdisciplinary understanding of material culture and technoculture.

Ecology

Ecology is the totality of relations among human agents, nonhuman agents, and their environment. Understanding material culture and technoculture as ecology means focusing on complex patterns of relationships rather than linear, hierarchical, and atomistic causes and effects (see Couch 1995). Relatedly, understanding material culture and technoculture as ecology allows one to dismiss facile views of technological change as "happy pastorals of progress or grim narratives of power and domination" (Carey 1989:9). An ecological way of conceptualizing their subject matter is central to pragmatism and symbolic interactionism (e.g., see Carey 1999), performance theory (in particular, see Schechner's [2005:113-169] concept of ecological rituals 113-169), and the Gibson-inspired (1979) view of social semiotics espoused by Van Leeuwen (2005). In this section I wish to outline three interdependent components of this ecological view: a temporal one, a spatial one, and the one based on self-indication and negotiation.

To exemplify these ideas I am going to refer to my ongoing ethnographic study of ferry boat mobility on the west coast of British Columbia. To understand the cultural significance of a ferry boat to an island community we can think of the relation of the ferry to time, space, and the self-indications of the people who depend on it. A ferry's daily and weekly schedule, navigation speed, and the duration of its travel to destinations deeply shape islanders' sense of time, including hourly (see Hodson and Vannini 2007), daily, weekly, and seasonal rhythms. Changes in scheduling are consequential. Changes in seasonal schedules made to accommodate higher tourist travel demand result in faster-paced life in the island during summer and in the consequent temporary disintegration of some social bonds and rituals. For example, during the slow seasons of autumn, winter, and spring, on Thetis Island it is customary to wave at oncoming traffic and pedestrians while driving. But because of the higher intensity of foot and car traffic this ritual is suspended during summer: "we wave at drivers during the slow seasons even before we recognize who is driving because chances are we know that person"—one of the 300 residents of Thetis Island explained to me—"but what's the point at waving in the summer when chances are that you either don't know the other driver, or more importantly that they don't understand the courtesy rule of actually waving back?"

A ferry is also enmeshed in spatial relations (see Vannini and Vannini forthcoming). The carrying and loading capacity of a ferry, for example, deeply shapes the space of islands and the sense of place of their residents. The relatively recent innovations in RORO (Roll On-board, Roll Off-board) technology have made the slow and risky boat side-loading of cars and freight antiquate, car-commuting via ferry faster, safer, and more practical, and consequently introduced small and remote islands to the road infrastructure and mobility culture typical of an automobile-based society. Of course this innovation is not the direct *cause* of changes, for these innovations merely introduced a possibility or an affordance that islanders interpreted and negotiated. Thus, self-indications do matter greatly. A self-indication is a process whereby persons point out to themselves the meaning of signs and consequences of actions. Some of the island communities I study, for instance, have rejected car ferries (but not foot passenger-only ones) entirely or seriously limited their sailing frequency, whereas others have either fully embraced them or rejected ferry service altogether. An ecological understanding of these dynamics requires that one take into consideration the multiple agents involved in the relationships that make up a technoculture as well as their self-indications and negotiations.

Ecology is not a structure but an ongoing process. Ecological perspectives need be sensitive to change, adaptation, integration, reintegration, and disintegration. Relationships among human, nonhuman, and environmental actors afford (cf. Gibson 1979) polysemic and multifunctional opportunities for the constitution of new meanings, signs, interpretations, technics, uses, and techniques. Material culture and technoculture are therefore "the life-energy holding together diverse symbolic and material organisms" tending toward movement and change (Vannini, Hodson, and Vannini forthcoming). At the ontological core of this life-energy tension are two dynamics marking all ecological settings: integration and disintegration. In the limited space remaining here I want to mention two shapes that integration and disintegration can take: *ecological carving* and *material claims-making* (for more see Vannini, Hodson, and Vannini forthcoming).

Ecological carving is a concept that describes action oriented at "carv[ing] out ecological niches" (Carey 1999:90).[1] Carving occurs by transforming environmental, material, symbolic, and human resources with the purpose of parceling out space and time and manipulating the boundaries and relations of these and other resources. The carving metaphor is intended to simultaneously convey the symbolic and material (though these concepts, as said, are neither opposed nor different) aspects of technoculture and material culture. The metaphor also highlights the orientation to bodily practice and to the everyday physical and material engagement of the world. As an expressive and instrumental action, the practice of carving assembles preexisting elements of, and potential for, form while engaging in creative transformation.

Interestingly enough, a carved object is itself a tool that works not only as a representation of the world, but also as a ready-at-hand technic that can be used for future carving and transformation. And it is no passive instrument either. Just like by carving the world humans make a claim on their possession of it, the nonhuman world engages in material claims-making of its own. Making a claim means to ask for something, to require something, and at times even to take something. This dramatic process is at the very basis of ecology. For instance, in "carving" a new scheduled ferry route to and from a remote island, islanders claim water as a space for making connections with the outside world. But the material world is no inert matter, offering resistance (Mead 1938), demanding readaptation, and making claims of its own. For instance, for its effective operation a ferry asks for an infrastructure that radically alters the insularity of an island, requires regular patronage by its users (lest services be too expensive or cancelled due to the economic unfeasibility of infrequent travel), and even takes things away from its users (e.g., the freedom to structure time regardless of schedules, the isolated character of island spaces, etc.). Making a claim is not synonymous with independent determination. A claim must be answered, considered, and can even be resisted and rejected. A claim is answered through self-indication, negotiation, and further ecological carving. The dramatic iterativity of carving and material claims-making—and the consequent drama of integration, disintegration, and reintegration (see Pfaffenberger 1992; Turner 1988)—shifts emphasis away from the idea of ecology as mechanical equilibrium (see Carey 1999:92) and exposes the emergent character of ecology.

Diffused Agency

Agency is "the socio-culturally mediated capacity to act" (Ahern 2001:110). As discussed throughout this book, anything—human or nonhuman, alive or not—has capacity for action. The principle of diffused (rather than merely human) agency underlines the active character of all subjects and objects implicated in material culture and technoculture contexts. Anything from individuals to single technological devices or natural objects (McCarthy 1984), and from persons and groups to whole infrastructures (Star 1999) can be directly involved in the making of culture. Extending conceptualizations of agency from humans to nonhuman actors is useful and helpful because it inductively parallels, describes, and captures the common, everyday processes through which humans *act as if* the objects they interact with *had* agency. Gell (1998) brilliantly demonstrates this through the example of a little girl playing with a doll and acting toward it as if the doll had human feelings, and the example of adults acting toward religious indexes, icons, and gods as if they had an obvious capacity for action. Arguably, neither plastic dolls nor photographic reproductions of gods (or live, embodied idols themselves) can act on their own, but in settings where significance and force is attributed to them, their agency clearly transpires as an impression not all different in kind or intensity from the impression that humans have of their own agency. Thus, while it attempts to make sense of anthropomorphism, the principle of diffused agency dismisses all forms of determinism—from technological and material determinism, to biological and social reductionism—as erroneous. Understood this way agency is thus not something that a human being has, but is instead the diffused potential for action present in a social and material setting.

One of the fundamental contributions of material culture and technoculture studies is the idea of mutuality. While contemporary symbolic interactionists have been more or less resistant toward this notion (though see the recent intervention by Owens [2007]), pragmatists and symbolic anthropologists have long agreed that "the lines between persons and things are culturally variable...[and that] in certain contexts persons can seem to take on the attributes of things and things can seem to act almost as persons" (Hoskins 2006:74). Not only does this notion open itself to the idea that nonhuman things have biographies (see Woodward this volume), but it also lends itself to the characterization that things play a key role in everyday social dramas. Performance theory (e.g., Schechner 2005) and performance-based analytical approaches to the study of material culture and technoculture (see J. Mitchell 2006) are perfectly well suited for the study of material culture and technoculture as interaction.

As the etymology of the word suggests, drama is about doing, action. Performance-centered approaches can emphasize the value of studying the inter*action*—of both the human-human and the human-not human types—that lies at the basis of rituals, play, ceremonies, spectacles, crises, social dramas, cultural performances, and other everyday life performances. Performance-based approaches are in the position to highlight how in its performativity diffused agency is relative across contexts, dependent on relationships between actors, and how it is primarily concerned with creation, process, and practice.

Objects perform. That is why we commonly refer to such things as high-performance vehicles, athletic performance shoes and gear, and so forth. Of course they do not perform alone. But neither do people; people and things perform together. We can think of material objects not as given or essences, but in light of *how* they perform and *what* they perform (i.e., the scripts they are endowed with and enact). For example, as I have shown in my recent ethnographic writing on the sinking of the ferry *Queen of the North* (Vannini 2008), this boat's unique relationships with her passengers over time, her reliable performance at sea, and her painful demise created a powerfully convincing impression of her personhood and agency. As the ferry sank and people began to mourn, a clear *performative transformation* occurred: right as the boat ceased to be functional as a machine she became more and more of a meaningful person to the residents of the British Columbia coast.

The idea of performative transformation—which the sinking of the *Queen of the North* epitomizes—is a central one in both performance studies and material culture and technoculture studies. Simply put, performative transformation is an exercise in agency, and more specifically a creative passage entailing change in form and subjectivity. For example, it is through performative transformation that material objects become social subjects (see Gell 1998; Kuechler 2002). As John Mitchell (2006) points out in his review of performance-based approaches to material culture, performative transformation has as its focus the body, space, and things. Typical instances of performative transformation are initiation rites, death rituals (the *Queen of the North*'s sinking being one of them), masquerading, and various other more or less extraordinary cultural performances and mundane social dramas (for an overview of what performances may entail see Schechner 2005). In relation to social dramas following the theoretical lead of Turner (1988), Pfaffenberger (1992) has usefully conceptualized how conflicts arising over technological adaptation, innovation, and change work as the chronological succession of breach, crisis, redress, and reintegration—an outline I also adopted in the ethnographic study of the *Queen of the North*'s sinking.

This treatment of drama and performance aims to transcend, while capitalizing on, Goffmanian dramaturgy and its limited attention to objects as inanimate "props" for impression management. Furthermore, it intends to show that diffused agency is not some kind of a stock resource that one has, but is instead both potential/capacity for action and its actualization. Just like gender is not something that one has, but is instead something that one does, diffused agency is not something that material technics have, but is instead something based on performance. Another way of saying this is the true characteristic of materiality is not its essence, but instead its consequentiality, thus its agency.

In giving heightened attention to the agency of material objects it is important not to lose sight of how agency is *diffused*. An overpreoccupation with the agency of objects might reintroduce into material culture and technoculture studies a form of technological or material determinism in disguise or perhaps even an instance of animism. Rather than agency alone and on wherein it lies, therefore, it is best to focus ethnographic attention on the dramatic and creative ways of relating between humans and nonhumans (see Gell 1998). In this sense, to speak of diffused agency is to invoke an ecology of interaction.

Emergence

As described by Mead (1938:641) the concept of emergence refers to a process marked by interactive indeterminacy: "when things get together, there arises something that was not there before." Such a seemingly simple idea is at the core of the perspective brought forth here as it outlines how interaction outcomes are inevitably conditional on both the action of all the actors involved and the nature of the preexisting conditions in which action takes place. Conditionality is a fundamental characteristic of interaction: "once an emergent is formed through a particular process of interaction, it acquires certain characteristics that are qualitatively different from those of the preexisting conditions on the basis of which this interaction has taken place" (Chang 2004:413). An emergent, therefore, is the offspring of the enabling and inhibiting elements of the actions of all those involved. No determinism is possible.

In Mead's (1938) philosophy of the act we can find another key property of the idea of emergence. The process of the act—and its four phases of impulse, perception, manipulation, and consummation—depend at all times on the conditioning that the past exerts on the present. "Each phase," Chang (2004:408) explains in his interpretation of Mead's thought, "is constituted by a specific pattern of interrelation between the actor and its environment that generates certain kinds of emergents. Many of these emergents act back on the process of interaction as mediating factors." Such a focus on the temporality of action and on historical trajectories is central in much of the interactionist (and constructionist, as we have seen in chapter three) research on technoscience (for a comprehensive review of this field see Clarke and Star 2003).

Technoscience studies lie at the intersection of the history of technology, the interdisciplinary study of information technologies, and at least five different but closely related sociological fields: the study of knowledge, of work, of science, of medicine, and of technology.[2] Ethnographic studies in this vast field have focused on work practices, materials, and what most famously have come to be known as boundary objects (Star and Griesemer 1989); classification systems (e.g., see Bowker and Star 1999); clinical research and trials; knowledge and technology in the biomedical sciences (e.g., see Star 1995); computing and information technologies in the workplace (e.g., see Orr 1996); the constitution of scientific and technological disciplines and specialties, as well as the role played by controversies in the shaping of these social worlds (e.g., see Clarke 1998).

An important manifestation of the principle of emergence in material culture and technoculture is the process of *semiotic transformation*. The concept of semiotic transformation refers to change intervening over time (i.e., diachronically) in the way objects are used and/or in the meaning attributed to signs. For instance, in my study of the culture of artificial suntanning (Vannini and McCright 2004) Aaron McCright and I observed how the meanings of tanned skin have varied over time and how both the sun and its delegated technics (namely, the artificial tanning lamp) have been put to different uses over time. Going back to the nineteenth century and early twentieth century, for example, we noted how in North America tanned skin connoted humble class origins due to its associations with outdoor work culture such as farming. Later—with the growing importance of outdoor leisure and tourism for the cultivation of individual lifestyle and taste distinction—tanned skin began to connote health, sociability, and class status, and thus to be viewed as attractive. Concurrently, medical authorities became more vocal about the benefits of prolonged exposure of the skin to the sun's UV rays. Later in the 1980s and 1990s the invention and domestication (in the Western world) of the tanning lamp enabled those who could not afford to take periodical vacations to sunny locations to conspicuously display their bronzed skin and, implicitly, make status and identity claims. But with growing evidence in the late 1990s of the carcinogenic and otherwise harmful effects of excessive exposure to UV rays, and the growing sensitivity of the public to these issues, tanned skin began to be increasingly associated with poor health and tanning lamps with danger. Arguably as a response to such bad press, much of the artificial tanning industry today promotes tanning lamps as effective devices to combat the winter blues, or what in commercial/medical terms has come to be known as seasonal affective disorder. Semiotic transformations are deeply interconnected with the power of meaning, and the power to frame and influence meaning: the topic of the next section.

Semiotic Power

Semiotic power (cf. Wiley 1994) refers to the consequentiality of representation. As Keane (1997:7) has remarked, representation is to be intended as both "depiction (representation *as* something) and delegation (representation *by* someone or something." In other words, representation may work as something that "addresses itself to a mind" (Peirce 1986:62), but it may also work as something that has "qualities independent of its meaning" and something that has a "causal connection with its object" (62).

To explain these ideas more clearly let us take the example of the human body and its capacity for sensation. In a series of qualitative investigations Dennis Waskul and I (Waskul and Vannini 2008; Waskul, Vannini, and Wiesen 2007) have begun to question the principle of *symboli*zation (cf. Snow 2001). As Rochberg-Halton (1982), Knappett (2005), Gottdiener (1995), Dant (2005), and Keane (1997, 2003) have remarked, to think of symbolization is to disregard how representation does not always take place through symbols. Bodily sensations, for example, are often meaningful without having to rely on symbolism for their functioning. Take, for example, sexual/sensual touch. Stimulating one's own body or others' sexually is not arousing because of, or thanks to, its abstract (e.g., discursive) properties. Surely those types of mental associations may play a role, but tactile sensations are first meaningful at a more basic level: the level of their qualitative immediacy (cf. Dewey 1934) or firstness (Peirce 1986). Thus, masturbatory stimulation may be meaningful well before an individual has acquired the symbols (words, discourses, values, etc.) to define the social significance of this practice (Waskul, Vannini, and Wiesen 2007).

What the above example highlights is that at the very least, according to Peircean semiotics or social semiotics (Van Leeuwen 2005; Vannini 2007), representation is of three types: iconic, indexical, and symbolic. Symbols—like words, or a stop light at an intersection, or a diamond ring—can be "read" effectively with the meaning intended by the producer only after an individual has been socialized to the semiotic conventions and lexical rules of a culture. Symbolic representation is the most commonly examined kind of representation—especially due to its relevance in structuralist and post-structuralist research and theorizing—as it highlights a fundamental characteristic of human interaction: the ability to create signs and conventions on which all members of a culture rely for finding meaning. The study of symbolic representation has been so prevalent in the social and cultural sciences that many scholars have erroneously begun to equate semiotics with the study of symbolism. And because many students of symbolism have a tendency to view symbols as ephemeral, deeply rooted in discursive practices, extremely malleable to the vagaries of social construction and deconstruction, and entirely dependent on mental associations and social conventions for their meaning, several scholars of material culture have begun to dismiss the very usefulness of semiotics. But this need not be so, for other kinds of representation are extremely important to a materialist version of material cultural studies, namely indexicality and iconicity.

Iconicity is a relation of continuity, self-reference, or resemblance between an object and the form its representation takes. Realistic drawings and sculpture are a classic example of this. Indexicality on the other hand is a relation between action and its consequences. Lightning is indexical of a storm, for example, or the olfactory sensation of a smell indicates that an odoriferous source must have created it. Both iconicity and indexicality require interpretation like symbolism does, but the interpretation of the former is rather different in nature because it deeply relies on habit, skill, and taken-for-granted relations. Because in everyday life they most often go unquestioned it is possible to say that iconic and indexical meanings are particularly powerful.

This is the case of bad smells, for example. As we found after asking a diverse group of people through the use of research diaries, very often what smells bad is viewed as bad—that is, morally despicable or aesthetically inferior—and treated as such. Thus a foul-smelling home is uncivilized, a foul-smelling individual is rude (or worse), and fragrance is the aroma of sanctity, purity, beauty, or health (Waskul and Vannini 2008). What is even more interesting is that these relations are immediate; we do not just smell an odor, carefully debate its possible values, and then select the most appropriate set of denotations and connotations. Rather, we smell something bad, unhealthy, low class (Waskul and Vannini 2008), and simultaneously engage in instantaneous self-indication. Semiotic approaches to the materiality of everyday life like this highlight how the material world has an immediate qualitative potential for meaning-making (Dewey 1934) and how it exercises consequential action.

In arguing for semiotic power instead of symbolism I am therefore arguing for a tripartite idea of the sign and representation that overlaps with the idea of techne (Vannini, Hodson, and Vannini forthcoming). As Vannini, Hodson, and Vannini discuss,

> According to Peircean pragmatist semiotics meaning emerges out of the triadic interaction between an object, a sign vehicle—known as representamen—that stands for that object, and the sense that someone makes of this relation—known as interpretant. This triadic model of signification can be extended to techne. A basic instance of techne—that is, a single technological act—can be explained through a similar model which comprises also three units: a technic, a technique, and an object to which the technique and technic are directed. The elements within the models of semiosis and techne overlap as follows: the representamen corresponds to the technic, the interpretant to the technique, and the semiotic object in Peirce's model corresponds to the desired end to which the technic-mediated technique is directed. Within this model interpretation and embodied practice, meaning and bodily purpose, symbolic and material mediation of the world overlap.

Understanding techne and representation as the sides of the same coin allows us to view signification and delegation through the same lens. This is a lens that frames action and in particular interaction with material objects as having a "social dimension beyond their symbolic meaning" (Dant 2005:111) and this suggests that, as Dant (111, my emphasis) continues, "as the social human being interacts with an object, she or he must take account of what the object is *doing* or about to *do* and must fit their line of *activity* to the intentions imbedded in the object." Thus, in doing something with objects, in embodied action, in the practical undertaking of the materiality of the world we tend to rely less on symbolism and more on the power that objects afford. Understood this way these are signs and "things that shape the self and the mind" as well as the body (Carey 1989:316). And finally, understood this way material culture, technoculture, interaction, and culture become inseparable and synonymous.

Summary

Interactionist approaches to material technoculture are based on four intersecting principles: ecology, diffused agency, emergence, and semiotic power. Together, these principles demonstrate how material technoculture resides neither in nonhuman objects nor in human actors, but instead in the emergent product of their interaction. The application of these principles to empirical research design and data analysis should free researchers from imposing either humanistic lenses on the material world (which impose an excessively *ethnos*-centric ideology on data) or deterministic ones (which impose reductionist tendencies). Beside their potential for postdualist material culture studies, the application of these four principles has also the obvious potential of changing ethnography as a strategy of data collection, analysis, and representation.

Notes

1. On the significance of carving as metaphor see also Tilley (2002). On the related significance of bricolage see R. Mitchell (2000) and on weaving as metaphor see Ingold (2000).
2. Though technoscience studies have much in common, at least theoretically, with material culture and technoculture studies, their empirical concern is often different.

References

Ahern, Laura. 2001. "Language and Agency." *Annual Review of Anthropology*, 30:109–137.
Becker, Howard. 1986. *Doing Things Together: Selected Essays*. Chicago: Northwestern University Press.
Bowker, Geoffrey and Susan Leigh Star. 1999. *Sorting Things Out: Classification and Its Consequences*. Cambridge, MA: MIT Press.
Carey, James. 1989. *Communication as Culture: Essays on Media and Society*. New York: Routledge.
———. 1999. "Innis 'in' Chicago: Hope as the Dire of Discovery." Pp. 81–104 in *Harold Innis in the New Century: Reflections and Refractions*, edited by C. Acland and W. Buxton. Montreal and Kingston: McGill-Queen's University Press.
Chang, Johannes Han-Yin. 2004. "Mead's Theory of Emergence as a Framework for Multilevel Sociological Inquiry." *Symbolic Interaction*, 4:405–427.
Clarke, Adele. 1998. *Disciplining Reproduction: Modernity, American Life Sciences and the "Problems of Sex."* Berkeley: University of California Press.

Clarke, Adele and Susan Leigh Star. 2003. "Science, Technology, and Medicine Studies." Pp. 539-574 in *The Handbook of Symbolic Interactionism*, edited by Larry Reynolds and Nancy Herman-Kinney. Thousand Oaks, CA: Sage.
Couch, Carl. 1995. "Oh, What Webs Those Phantoms Spin." *Symbolic Interaction*, 18:229-245.
Dant, Tim. 2005. *Materiality and Society.* New York: McGraw-Hill.
Dewey, John. 1934. *Art as Experience.* New York: Capricorn.
Gell, Alfred. 1998. *Art and Agency.* Oxford: Oxford University Press.
Gibson, James. 1979. *The Ecological Approach to Visual Perception.* Boston, MA: Houghton Mifflin.
Gottdiener, Mark. 1995. *Postmodern Semiotics: Material Culture and the Forms of Postmodern Life.* New York: Blackwell.
Hodson, Jaigris and Phillip Vannini. 2007. "Island Time: The Media Logic and Ritual of Ferry Commuting on Gabriola Island, BC." *Canadian Journal of Communication*, 32:261-275.
Hoskins, Janet. 2006. "Agency, Biography, and Objects." Pp. 74-84 in *The Handbook of Material Culture*, edited by Chris Tilley et al. London: Sage.
Ingold, Timothy. 2000. "Making Culture and Weaving the World." Pp. 50-71 in *Matter, Materiality, and Modern Culture*, edited by Paul Graves-Brown. New York: Routledge.
Keane, Webb. 1997. *Signs of Recognition.* Berkeley: University of California Press.
——. 2003. "Semiotics and the Social Analysis of Material Things." *Language and Communication*, 23:409-425.
Knappett, Carl. 2005. *Thinking through Material Culture: An Interdisciplinary Perspective.* Philadelphia: University of Pennsylvania Press.
Kuechler, Susanne. 2002. *Malanggan: Art, Memory, and Sacrifice.* Oxford: Berg.
McCarthy, E. Doyle. 1984. "Toward a Sociology of the Physical World: George Herbert Mead on Physical Objects." *Studies in Symbolic Interaction*, 5:105-121.
Mead, George. 1938. *The Philosophy of the Act.* Chicago: University of Chicago Press.
Mitchell, John. 2006. "Performance." Pp. 384-401 in *The Handbook of Material Culture*, edited by Chris Tilley et al. London: Sage.
Mitchell, Richard Jr. 2000. *Dancing at Armageddon: Survivalism and Chaos in Modern Times.* Chicago: University of Chicago Press.
Orr, Julian. 1996. *Thinking about Machines: An Ethnography of a Modern Job.* New York: ILR Press.
Owens, Erica. 2007. "Non-biologic Objects as Actors." *Symbolic Interaction*, 30:567-584.
Peirce, Charles. 1986. *Writings of Charles Sanders Peirce*, vol. 3, edited by M. Fisch et al. Bloomington: Indiana University Press.
Pfaffenberger, Bryan. 1992. "Technological Dramas." *Science, Technology, and Human Values*, 17:282-312.
Rochberg-Halton, Eugene. 1982. "Situation, Structure, and the Context of Meaning." *Sociological Quarterly*, 23:455-476.
Schechner, Richard. 2005. *Performance Theory.* New York: Routledge.
Snow, David. 2001. "Extending and Broadening Blumer's Conceptualization of Symbolic Interactionism." *Symbolic Interaction*, 24:367-377.
Star, Susan Leigh. 1995. *Ecologies of Knowledge: Work and Politics in Science and Technology.* Albany, NY: SUNY Press.
——. 1999. "The Ethnography of Infrastructure." *American Behavioral Scientist*, 43:377-391.
Star, Susan Leigh and James R. Griesemer. 1989. "Institutional Ecology, 'Translations,' and Boundary Objects: Amateurs and Professionals in Berkeley's Museum of Vertebrate

Zoology, 1907-1939." *Social Studies of Science,* 19:387-420.
Tilley, Chris. 2002. "The Metaphorical Transformations of Wala Canoes." Pp. 63-80 in *The Material Culture Reader,* edited by Victor Buchli. Oxford: Berg.
Turner, Victor. 1988. *The Anthropology of Performance.* New York: PAJ.
Van Leeuwen, Theo. 2005. *Introducing Social Semiotics.* London: Routledge.
Vannini, Phillip. 2007. "Social Semiotics and Fieldwork: Method and Analytics." *Qualitative Inquiry,* 13:113-140.
———. 2008. "A Queen's Drowning: Material Culture and the Drama of a Technological Accident." *Symbolic Interaction,* 31:155-182.
Vannini, Phillip, Jaigris Hodson, and April Vannini. Forthcoming. "Toward a Technography of Everyday Life: The Methodological Legacy of James W. Carey's Ecology of Technoculture as Communication. *Cultural Studies ← → Critical Methodologies.*
Vannini, Phillip and Aaron McCright. 2004. "To Die For: The Semiotic Seductive Power of the Tanned Body." *Symbolic Interaction,* 27:309-332.
Vannini, Phillip and April Vannini. 2009. "Of Walking Shoes, Boats, Golf Carts, Bicycles, and a Slow Technoculture: A Technography of Movement and Embodied Media on Protection Island, BC." *Qualitative Inquiry,* 14:1272-1301.
Waskul, Dennis and Phillip Vannini. 2008. "Smell, Odor, and Somatic Work: Sense-Making and Sensory Management." *Social Psychology Quarterly,* 71:53-71.
Waskul, Dennis, Phillip Vannini, and Desiree Wiesen. 2007. "Women and Their Clitoris: Personal Discovery, Signification, and Use." *Symbolic Interaction,* 30:151-174.
Wiley, Norbert. 1994. *The Semiotic Self.* Chicago: University of Chicago Press.

PART 2
ETHNOGRAPHIC STRATEGIES OF REPRESENTINGTHE MATERIAL WORLD

6
From Embodied Ethnography to the Anthropology of Material Culture: Gaming in the Field

Mélanie Roustan

The position and uses of the body not only as an object of study but also as a tool of knowledge—for informants as well as for ethnographers—continue to feature at the center of metatheoretical and methodological reflections across disciplines and fields in the social sciences. As debates rage on, the conditions and possibilities of "embodied ethnography" as a research strategy, and even more, as an epistemology, are at stake. In particular, how can prediscursive capacities be studied and used as sociocultural competency? To what extent can bodily apprenticeship be considered as a mode of knowledge transmission and technique for social inquiry? The recent debate on *Body and Soul* (*Qualitative Sociology* 2005; *Symbolic Interaction* 2005; Wacquant 2004) exemplifies this ongoing reflection in the context of an evolving "carnal sociology" that claims to be not only "of the body (as a social product)" but also "from the body (as social spring and vector of knowledge)" (Wacquant 2005). In anthropology, these kinds of reflection are to be found in the material culture paradigm (e.g., in the French context see Julien and Rosselin 2005; Roustan 2007; Warnier 1999). Focusing on objects, this paradigm shows a constant preoccupation to avoid considering bodies as "dead things" (even in archaeology or museology) but instead as parts of human action, techniques and habits, and then to situate the body at the core of the processes of socialization and subjectivization.

The purpose of this chapter is to offer a reflection on these issues in the context of the narration of my experience as an anthropologist on a fieldwork project on computer and video games (Roustan 2003, 2007).[1] The aim is to confront technology through embodied ethnography and a material culture paradigm.

Several levels of analysis will intersect with each other. Personal methodological reflection, field experience, and critical thinking will come to sustain more general epistemological remarks on ethnographic practices, especially concerning the uses of the ethnographer's body as a tool of research (in a technological context). Simultaneously, my aim will be to question technology as a specific research object in anthropology. The study of material culture will be considered a category of ethnographic research centered on objects and their uses, as well as a way of doing anthropology by focusing on the physical dimensions of subjective experience. In addition to my own body as a heuristic instrument, the bodies of my informants will also become central to my reflection on data gathering and conceptualization. Furthermore, I will reflect on the following: when it comes to embodied ethnography, what and whose bodies are at stake? How can such a dialogue "between bodies" work and be productive in terms of knowledge, especially on technology? On which postulates does it sit and what kind of mechanisms does it suggest?

"Dematerialization" as a Discourse on Technology

Representations of "new technologies" such as games are polarized between two imaginaries: a "messianic" one and an "apocalyptic" one. For some commentators—especially journalists, but also social scientists and writers—the spread of new technologies is like a blessing for human beings. According to them new technologies will enhance human capacities and abilities in such a way that the great technical and philosophical questions of our times will be solved. For other commentators at the opposite end of the spectrum, new media and tools are going to lead the world to its end because technology will surpass human beings and maybe dominate them or even replace them, or at least alienate them. In both messianic and apocalyptic discourses the focus is on what is going to occur in the future and in a prospective rhetoric on the uses, utility, and dangers of such tools. In both discourses the debate is also organized around similar characteristics of these emerging phenomena: their influence on human interactions, communications, and identifications, their impact on production and circulation of information, and on the distribution of power. In doing so, these discourses tend to position technology as a cause of social changes, thus somewhat neglecting to consider how technology itself is the product and outcome of social organization (see Pinch this volume).

One can also easily identify a transversal axis in those public discourses. This is an axis that finds its semantic concretization in such words as "dematerialization" or "virtual." In other words, seemingly those mysterious machines change not only society but also our relationships to *reality*.

On the other hand, such esoteric, ethereal attitudes, and conclusions are hardly ever typical of ethnographic research. Fieldwork—even in such fields of study as new technologies—usually begins with practical concerns and physical interaction; necessary steps to organize "indigenous encounters." Thus, phone calls must be made and words have to be exchanged to allow and engage face-to-face meetings. Then, information is usually gathered during informal or formal face-to-face or voice-to-voice discussions or interviews with informants. And all of this feels hardly dematerialized or virtual. In sharp contrast with this, when I started my fieldwork on gaming, I met some players hoping they could shed some light on this topic for me and introduce me to some playing situations in which they or their friends and/or competitors were involved. As we began to talk our early exchanges remained at an abstract and "theoretical" level (as opposite to the "practical" feel of our interaction) that dealt with their awareness of their stigma as players: a stigma based on the stupefying and addictive reputation of their practices. Our conversations also revealed a larger appropriation of social discourses on virtuality in either its magic and mystique and its potential for harmfulness. In other words, despite the all-but-virtual feel of our interaction our discourses tended to stay at the level of representation and generality typical of those who hold new technologies in utopian or dystopian reverence. Indeed a classic starting point for discussions consisted in taking for granted that computers and electronic tools let people accede to a less concrete and material world than the one they are used to. With such new fancy stuff, they are supposed to be able to discover new evanescent universes, either for better or for worse. These ideas tended to be unquestioned at the beginning of our interactions. And that is why it seemed important to me right away to focus on facts rather than on opinions or more precisely to encourage people to describe practices in great detail, as well as their feelings and judgments, and the qualities of material objects, times, spaces, techniques, interactions, and so on. During discussions, representations always come alongside with practices, which is not true on a reciprocal basis (discourses may focus on representations without giving information on practices).

Even though fieldwork research is supposed to be inductive and the ethnographer somewhat neutral (which are just myths), it looks like questions always do influence answers. It is indeed part of a researcher's role to bring out specific topics over debate. And so I did. When I started to encourage discourse on the materiality of their practices, gamers appeared very prone to talk about their machines: their weight, their size, their reliability or fragility, their cost, and so on; in fact, one just needed to ask! Lots of stories seemed available on how to obtain this or that hardware or software—on how the "system of provision" (Fine and Leopold 1993) was working. With talk on the specific objects and actions required by the activity there came information on the technical skills they needed to work, in the double sense of staying in service (maintenance) and fulfilling their function and thus giving players satisfaction. Consequently, the body emerged as a main topic of conversation: its pleasures (the joy of control and performance, the excitement of bodily projection on the screen); its constraints and resistances (learning, trying again and again); its natural needs (eating, drinking, sleeping, going to the bathroom); sometimes its pains (exhaustions, cramps, or tendonitis). As usual, it is when it "fails" that the body is noticed and can be put in words, obviously not when it disappears in routine. Therefore, behind a thin slice of "prêt-à-parler"[2] based on easily available social discourses on games as parts of a new dematerialized world, there laid more personal narratives on practices, material objects, the body's involvement, and interactions with other people.

As this transpired, a first methodological lesson presented itself: if ethnography presupposes a comprehensive approach to local points of view, it does not mean that interactions during fieldwork escape from general rules on the construction of discourses—among them the fact that interviewers' questions largely condition the content of answers. There is no use considering this phenomenon as a bias on data gathering; interaction lies at the core of ethnography and must be taken into epistemological account like any other element nourishing the debate on the researcher's influence on fieldwork. Indeed there is no pure or transparent relationship between the observer and the observed; both have strategic information to negotiate. Informants know that they are constructed as holders of information from the researcher's point of view (literally, as their name indicates), and vaguely understand that some part of their identity is at stake in the interaction. Ethnographers first appear as the ones who know nothing and ask for everything; then they start accumulating information until they have a lot of data on everything and everybody, including recipes or secrets, thus acquiring a potentially dominant position on a local scale (and maybe on a wider one with their writings).

Those who engage in ethnographic reflexivity have now been recognizing for a long time that the idea of neutral and even objective data gathering in contexts of human interaction is a pure fantasy. Indeed the social sciences in general have now refined description of mechanisms of discourse building among various situations and contexts and have acknowledged power relationships within fieldwork and representation (e.g., Clifford and Marcus 1986). It is thus worth working on this awareness of the researcher's embodied presence as a heuristic tool useful to define what is at stake in the phenomenon under study. This reflexivity is a sign of maturity for a discipline able to bridge the gap between science and literature in recognizing the importance of a researcher's subjectivity without abandoning objectification as a key process of scientific construction of results. So, how to do it?

Coming back to my own experience, the next step in "my" gaming fieldwork consisted in trying to achieve some kind of opportunity for participant observation. When I aimed to play, it was so hard just to manage to reach the starting point of their expertise: "where am I supposed to plug in this cord?" "Why do these machines seem to ignore each other even though they have been properly connected?" "How can my thumbs be so independent of my other fingers?" "And how am I supposed to manage to move them when my eyes and attention are turned to that screen?" Then, I quickly turned on two other thoughts. First—and for the best in terms of reassuring me about the potential dangers of this fieldwork—there was no risk that "virtual reality" could absorb me for the moment (it is too hard to reach!). Second, technology did not seem to constitute a particularly unique research object for an anthropologist: there are systems of material objects, representations, and beliefs, there are people impregnated with their socialization, exchanging tools, techniques, and views on things, trying to promote values and to negotiate norms of use and thinking, like for any other phenomenon I ever studied. So, during my early experience with this fieldwork, I drew on two directions for its development: a very specific one, following the mysterious road of so-called virtual reality—which appears at the same time to be central in discourses (as rhetoric) and in practice (as a goal to attain)—and a really unspecific one, leading me to consider gaming as a "classical" anthropological object and to treat it with my usual methodological tools of "classical" ethnographic fieldwork. These two orientations—desacralizing virtual reality as some magical thing and despecifying technology as a particular object in the context of material culture studies—are now to be deepened in what follows.

But before continuing, a remark has to be made on the idea of a "classical" object of anthropological research and about classical fieldwork in anthropology. Here I am not talking about a scientific tradition of the late nineteenth century or even of the first part of the twentieth. Rather, I refer to the new classical anthropology: the product of the recent but widely influential reflexive turn on doing ethnography in the context of the globalization of cultural flows (Appadurai 1996; Clifford 1997; Miller and Slater 2000; Warnier 2004). Like any contemporary object of study, the fieldwork that gaming implies is neither clearly geographically situated nor does it structure strict relationships of identification with the people it includes (Amit 1999; Marcus 1998). In other words it is not defined by a specific location on earth; it has to be considered as a combination of human and nonhuman elements and of material and immaterial contents; all factors that are constantly on the move. In the proper sense of movement—gamers can be either sitting at home or competing in international championships in Korea—technological affordances are constantly renewed and continuously offer innovative possibilities in terms of being here and/or there, of experiencing a sense of presence. Metaphorically, understanding movement also entails taking into account evolutions of technical supports and skills, reconfigurations of gaming as a child's hobby or a grown up profession, as a sport or a cultural good (or even as heritage collected in museums). Material aspects, cultural categorizations, and social functions of the practice are changing, and so fieldwork boundaries must adapt accordingly. Thus, what is at stake in this kind of ethnography is to grasp general dynamics of the phenomenon, as well as specific mechanisms of change and stabilization.

Virtual Reality: An Embodied Matter

One path I decided to follow for my research was that of virtual reality. Behind the mystique of the words, I wanted to understand what it was all about and how it worked. At this point, from an epistemological point of view, a few important points can be reiterated: ethnography is not (only) a semantic exercise; it cannot work only on discourses, even less on opinions, but it has to include concrete elements of people's life. Material objects are not to be reduced to mere signs (and social and cultural life as mere communication); they are to be considered as parts of a complex system of actions, as linked to human bodies in movement and then to subjects (on this see also Vannini chapter five).

With regard to the issue of the researcher's body I understood that if I really wanted to participate in the core activity in my field—interacting with games—I had to train hard and long, and only to reach a very minimal level of skill. And this was the first problem; as a researcher practicing "anthropology at home" I have specific obligations (such as teaching responsibilities, pressure on completing my PhD, etc.) and thus material and temporal constraints. The second and related problem was as I am older than 25 and have not been socialized to gaming since my early childhood, it is very hard, if not impossible, to manage to reach a sufficient level of skill to play with other gamers. This was obviously particularly true for games requiring body techniques. Even if I trained and trained again, I would always be an awkward gamer. This started resonating in specific excerpts from interviews and from discussions on how to gain expertise: female gamers are very few (at least for that first generation of gamers); usually, girls try to play during parties or other such social gatherings; they never reach boys' level of skill, and to a very large extent, women position themselves (according to men's testimonies) as against gaming, which they judge as stupid and time-consuming.

In sum, for the gamers I interacted with, the key laid in my femininity, especially in the way in which I have been socialized in the depth of my body techniques. So, as it turns out, it is the body and its dynamics—that is, the time needed to move it in space with objects, the process of taking on habits, routines, and technical expertise, modes of creating and actualizing meanings through action—that appear to be at the core of my questioning so-called virtual reality.

All of this shows me that virtual reality, if it exists, is not a given thing. It has to be performed, accomplished by working on one's body abilities to interact with a special machine till the point of appropriation of objects, of "incorporation of their dynamics" (Warnier 1999:34-37, 140-147), a point where movement becomes unconscious. It is not something powerful in the sense of a magical attraction that could wipe somebody off their feet. This constitutes a small but important step regarding the current opinion on games as dangerously powerful: bodily practices must be reintroduced in the reflection on gaming. Material objects cannot be thought outside of human actions that condition their production and their uses. The latter give them their reason of being: a function and some signification.

Virtual reality, considered at this stage of the reflection as what makes gaming specific and attractive, has a lot to do with the body and its senses and mechanisms of perception. The way people talk about it is centered on physical sensations: it is primarily based on a feeling of being and acting somewhere (in an imaginary universe represented on screen) and of being and acting elsewhere at the same time (in a sofa in front of a computer, in physical contact with a joystick or keyboard and eye contact with a screen). This double sensation of being here and there relies on a correspondence of action between the player's physical body and his/her avatar's virtual body. This possibility is offered by games but is not given: only a good player can produce it and experience it. The greater the body expertise is, the less attention is to be paid to it, like all commentators on "techniques of the body" (Mauss 1950[1934]), "habitus" (Bourdieu 1980), or "routine" (Giddens 1986) have noticed. Forgetting the mental effort necessary for action and taking recurrent gestures as unconscious action is what is at stake in the long time dedicated to train one's body to fluid action (indeed, I quickly understood why and how people tend to consider gaming as a sport). Action has a "double" meaning here, as I just said ("here" on my sofa and "there" on the screen; on this issue see also Tutt and Hindmarsh this volume; Richardson and Third this volume), and by this way generative of symbolic meanings. Games, like books, for instance (whose pages require to be turned to be read), combine action and signs, stratifying the potential of meaningfulness lying in human movement. With games, interactivity is reinforced as diverse alternatives can be activated at different times of the narrative (I fully understand why and how people also tend to consider and build games—as modes of expression of imaginary universes—as cultural goods).

Taking all this into consideration, trying to center data gathering on material culture means to focus on bodies (subjects) in action (not only on objects). Interview questions are to focus on body techniques, socialization, and expertise. They also have to take into account the rhetoric of virtual reality and its conditions of possibility when it comes to be experienced. At the observation level, it is critical to pay specific attention to bodily actions. Observing someone playing a car-racing game constitutes a good example. One can most easily admire the control and velocity of fingers on electronic tools, and recognize the precision of the union between eyes and hands in action (an apparent perfect appropriation of material objects by the player's body). But in addition, it is also possible to notice the way people move in front of the screen as if they were really driving a car, bending left or right as a driver would, contracting their neck and jaws when they are supposed to brake very quickly, tensing all their body during acceleration, and so on. Virtual reality is the result of interactivity between a machine (which splits and then deterritorializes action, with my fingers on a keyboard, my whole virtual body on the screen) and a human being, whose perceptions are organized as a multisensorial system.

No Objects without Subjects and Vice Versa

It is interesting to note that material objects never exist—as far as we can learn during interviews, discussions, or observation—out of a context defined by time, space, other objects, actions, and subjects. That is where material culture comes to a conceptual efficiency and how it finds its theoretical legitimacy. It is precisely because there cannot be any border between "techniques of the body," "techniques of instruments" (Mauss 1950[1934]), and "technologies of the self" (Foucault 2001) that it is worth doing material culture anthropology and more largely using embodied ethnography.

But let us go deeper into this point. To speak of material objects does not make sense if they are not considered as built and produced by human beings, on both concrete and metaphorical levels. Objects are concretizations of projects and uses. Even in the context of museums, for instance, when objects look frozen, when they appear as pure representations of culture and at the same time seem "autonomous" as objects, anthropology has shown that they are as socially and culturally constructed as in any other situation (Clifford 1988). The making of their status and significance results from a series of physical and symbolic actions directed to them (Julien and Rosselin 2005): from legal recognition as national heritage to restoration in workshops, from scientific and/or artistic classification to aesthetic arrangements among other objects (similar exhibits but also includes mediation tools). Objects are to be approached altogether with actions on them, and so with bodies in action, and thus with subjects.

Indeed, reciprocally, the body cannot be considered as dead flesh and envisaged without its apparel and the set of abilities it has been socialized to, through all the habits, routines, and rites, which give every subject technical skills and ways of acting and thinking his/her environment. At the individual level, it means that action on material objects is also an action on oneself. Let us now see to what extent it is also an action on other people's actions—in other words, some kind of power (Foucault 2001).

Gaming as Material Culture of the Everyday

After going back and forth between field and theory, and between data and analysis, a new scale of observation and interpretation for my fieldwork appeared to be pertinent. This scale, which would take into account gamers' environment, would add interactions (with human beings) to interactivity (with machines). This indeed became my second path: like for any other practice, the machine cannot be considered the *main* "actor." In addition, I realized, gaming cannot be reduced to the time and space of playing. A gamer is not only a man who interacts with a combination of hardware and software, he is also a student or/and a worker, a son, a brother, a boyfriend or a husband, a Parisian, a Brit, a French, a footballer, a philatelist, and so on. Practicing games cannot be considered as totally identifying a human being. Processes of identification are multiple and still only a part of the dynamics of becoming a subject.

In my case, the gender dimension appeared crucial. Themes and practices developed by games are key elements of "traditional" masculinity. The expertise required to access the gaming world demands an early bodily socialization to new technologies that seems to have been mostly given to boys, for that generation at least. So where are the girls? The first girl I found on the fieldwork was me! I finally found myself in that research position, not in the kind of participant observation I wanted to be in (a gamer among gamers) but one that was ascribed to me by the nature of the field (a young female researcher among gamers). There I was, among gamers, and through time had built up my position: sitting on the sofa, next to a player, and in front of the screen, screaming, congratulating, moving if necessary, going to the fridge and taking some food or beer; "Here we go, I am another girl in a Western domestic space!" Moving away from the sense of irony, I decided to widen my focus from videogaming practices to everyday life uses, and to gather ethnographic material on links between games and masculinity at a broader social level. Fieldwork and culture then appear as a complex web of significance (Geertz 1973) and as a series of objects, actions, and subjects, embedded into one another.

The time had come to decenter the focus, to move from a close scale to a farther one: from attention to interactivity between players' bodies and games as material and symbolic objects to a wider angle on the social interaction between people. Verbal communications with informants had then to be driven to encompass wider questions and so did observations. I had to take into account gaming as a source of leisure among available others putting at stake more social ones. Discussions on these topics underlined social norms and cultural frames structuring everyday life, and particularly gender relationships: private/public borders (online gaming blurring boundaries between public and private spheres, for instance, but also trivial questions of access to computer at home); leisure time partition (between work, household chores, personal enjoyment, and shared quality time with close relatives); choices of interactions (with colleagues, friends, competitors, and family, etc.); comparative legitimacies of objects, practices, and users (are games cultural goods or commercial products?), and so forth. In other words, questions of power at different levels of social life entered the so-called virtual world. Games had therefore to be considered from diverse points of view: as material objects in a large everyday life system of objects (technological or not); requiring actions on them (coming from subjects); implying actions on other people's actions; in other words an example of the circulation of power (cf. Foucault 2001).

Considering gaming as a system of interactions and focusing on the gender dimension led me to study two key issues: (1) gaming as medium of/for masculine sociability and socialization and (2) gaming as object of tension between men and women. These two "domains" are of course intimately intricate in the analysis but still remained independent at the stages of data gathering. When I came to pay particular attention to the various links between gaming and gender, it was obviously because of a common stereotype on the activity and, more concretely, of some reflexive remarks about my own capacity for integration in the field. The question of girls' and women's places in the gaming universe, on an everyday basis, interpellated me. My interactions and observations became directed mainly toward that direction, privileging mixed interaction contexts.

Three points can sum up this part of the fieldwork. As Vannini summarizes in chapter one, first and traditionally, women seem to be excluded from technical and technological worlds (it is difficult to meet female gamers recognized either as "real" gamer or "real" women) in a context where mastery of computers and systems of communication are a strategic element to reaching dominant positions. Second, gaming appears as a stumbling block between men and women (mother and son, girlfriend and boyfriend, wife and husband) in terms of domestic management, crystallizing tensions around household chore division, unequal access to leisure time, unequal legitimacy of gendered cultural consumption, and so on. Third, passionate relationships to games and girls seem to be inversely proportionate (the more you love, the less you play), leading to another source of tension when the activity tends to adopt the position of "mistress" in the relationship. Gaming plays an important part in the process of gender construction and performance as both instrumental to masculine sociability and as object of negotiation of rules and values across genders. Actions and interactions thus come together: the dynamics of material objects and of symbolic content (imaginaries, values, norms) are closely embedded. Circulation of objects conveys transaction of social and cultural values that leads to construction of a status for these elements, but also for people around them (Appadurai 1986).

In sum, from the starting point of my situatedness in the field I came to realize a crucial dimension of my research object. My realization on my situatedness shows that what is felt as reaching the limits of fieldwork for a given ethnographer, far from being barriers to his/her work, constitute the very uniqueness of his/her investigation and reflection. Participant observation, envisaged as the basic principle of the (embodied) ethnography, is first a tool based on reflexivity, and not so much useful for accumulation of positivist knowledge. Objectification of data relies on going back and forth between fieldwork and theory; it grounds itself on researcher's bodily perceptions and his/her subjectivity, on his/her feelings, and the way he/she has been welcomed and "placed" in fieldwork, including physically. In addition, it demonstrates how universal the questions are (like gender or relationships between action and signification—body and soul) that lead any empirical enquiry, whatever the forms they take and the contents they nourish. In the end technology does not arise as an object of research much more different or "virtual" than any other.

Notes

1. In the context of my doctoral research in anthropology, from 2000 to 2005 (University of Paris Descartes-Sorbonne), I met and followed twenty French video gamers (from occasional users to intensive ones), mainly young men.
2. Literally, "ready-to-talk," like "prêt-à-porter," means "ready-to-wear."

References

Amit, Vered (Ed.). 1999. *Constructing the Field: Ethnographic Fieldwork in the Contemporary World.* London: Routledge.

Appadurai, Arjun (Ed.). 1986. *The Social Life of Things: Commodities in Cultural Perspective.* New York: Cambridge University Press.

———. 1996. *Modernity at Large: Cultural Dimensions in Globalization.* Minneapolis, MN: University of Minnesota Press.

Bourdieu, Pierre. 1980. *Le Sens Pratique.* Paris: Minuit.

Clifford, James. 1988. *The Predicament of Culture: Twentieth Century Ethnography, Literature and Art.* Cambridge, MA: Harvard University Press.

———. 1997. *Routes: Travel and Translation in the Late Twentieth Century.* Cambridge, MA: Harvard University Press.

Clifford, James and George Marcus (Eds.). 1986. *Writing Culture: The Poetics and Politics of Ethnography.* Berkeley: University of California Press.

Fine, Ben and Ellen Leopold. 1993. *The World of Consumption.* London: Routledge.

Foucault, Michel. 2001. *Dits et Ecrits, Tome II: 1976–1988.* Paris: Gallimard.

Geertz, Clifford. 1973. *The Interpretation of Cultures.* New York: Basic Books.

Giddens, Anthony. 1986. *The Constitution of Society: Outline of a Theory of Structuration.* London: Polity Press.

Julien, Marie-Pierre and Céline Rosselin. 2005. *Culture Matérielle.* Paris: La Découverte.

Marcus, George. 1998. "Ethnography in/of the World System: The Emergence of Multi-sited Ethnography." Pp. 79–104 in *Ethnography through Thick and Thin*, edited by George Marcus. Princeton: Princeton University Press.

Mauss, Marcel. 1950 [1934]. *Sociologie et Anthropologie.* Paris: PUF.

Miller, Daniel and Don Slater. 2000. *The Internet: An Ethnographic Approach.* Oxford; New York: Berg.

Qualitative Sociology. 2005. Special Issue on *Body and Soul*, 28 (2).

Roustan, Mélanie (Ed.). 2003. *La Pratique du Jeu Vidéo: Réalité ou Virtualité?* Paris: L'Harmattan.

——— 2007. *Sous l'Emprise des Objets? Culture Matérielle et Autonomie.* Paris: L'Harmattan.

Symbolic Interaction. 2005. 28 (3).

Wacquant, Loïc. 2004. *Body and Soul: Notebooks of an Apprentice Boxer.* New York; Oxford: Oxford University Press.

———. 2005. "Carnal Connections: On Embodiment, Apprenticeship, and Membership." *Qualitative Sociology,* 28 (4):445–474.

Warnier, Jean-Pierre. 1999. *Construire la Culture Matérielle:* Paris: PUF.

———. 2004. *La Mondialisation de la Culture.* Paris: La Découverte.

7
On Driving a Car and Being a Family: An Autoethnography

Chaim Noy

Driving a car and riding inside one strike most of us as utterly mundane activities, about which little can be said and from which even less can be studied. Except for a number of scholars who have "everyday life" as their primary site of investigation, most look oddly at investigations of such routine activity as car-driving. This is certainly the case where driving is pursued in everyday urban settings, or settings that do not include any outstanding routes, events, or destinations. These eventless events are perhaps the epitome of the banality of routine urban life. Such are the trips I wish to discuss in this chapter. Specifically, the trips I want to discuss took place during the spring of 2006, when I would take my then-first-grade daughter, Noa, to her school. Although only few years have passed since, if I had not taken notes at that time, none of those mundane occurrences would have been remembered, or survived for reflection. This is true simply because there was nothing in particular to remember or to report about.

As Michel de Certeau (1984) famously argued, this quality of mundaneness, which is a consequence of the structure he called the "Ordinary," is nothing less than an ideological structure located at the political, moral, and ideological base of late modern life. In this structure, ordinary activity is unsigned, unreadable, and unsymbolized (xvii). If no one attends to it, it dissolves unattended, and we are gradually less and less informed about the practices and the environments that shape our lives, about the meanings they embody, and about the historical struggles in which we engage through them. Autoethnography, I argue, is a critical component of this endeavor.

Addressing motion itself as it is embodied in modern transportation systems is done in this chapter by examining the social practices and material settings of everyday car travel. In what follows, I argue that different social systems are embodied by and performed in the same place and in the course of the same interaction while driving. My aim, which is to show autoethnographically how multiple and situated social roles and meanings emerge in and through car-driving, will be accomplished by looking *inside the car* and by addressing how people *inside it* make use of the social possibilities and material affordances that are available by the car. This I accomplish through the sensitivities that reflexive and autoethnographic methods make available. Hopefully, with the help of reflexive methods I will be able to draw a *sensitive* portrayal of the inside of the car, that is, a portrayal that addresses the senses.

I argue that there are two social systems that emerge as relevant to this inquiry, which are juxtaposed in the events of taking my daughter to school. These social systems are the family, on the one hand, and transportation and specifically automobility, on the other. By referring to the family and the transportation as social systems, different roles, practices, meanings, spaces, and power relations are referred to. Because both systems are mutually informative, studying the interplay between them is illuminating. One can gain insights into families from studying cars and vice versa: the practices of driving can be understood when observing families. In fact, in what follows these systems emerge as enmeshed in each other, and are only analytically describable; in the reality of everyday life families and cars are mutually constitutive.

I use the term "system" following John Urry's (2000) influential thesis, concerning the "System of Automobility." According to Urry (2004:26), the automobility system is "an extraordinarily powerful *complex* constituted through technical and social interlinkages" (emphasis in original). Urry suggests a notion that is holistic, and that consists of multitude of sites, actors, spaces, practices, and representations, all of which are loosely (inter)connected, and comprise, together, a whole that is larger than its parts. While there are other possible conceptualizations of the term system, I find Urry's lead productive because it is conceptualized specifically in relation to transportation. In addition, Urry does not limit the system to "external" influence, but on the contrary, he necessitates linkages with other sociomaterial systems for its vitality and operation. Hence, it establishes a heterogeneity and multiplicity of interconnected and interconstitutive networks, which loosely link practices, meanings, and material objects and settings that might otherwise be rendered unrelated.

Autoethnography as a Tactic of Everyday Life

In the ensuing discussion I use two excerpts taken from my field notes, where I wrote my impressions of the motorized trips with my daughter to school. I use reflexive autoethnographic methods; methods where the observer is also an actor in the observed scene, for a number of reasons. First, reflexive methods allow us access to knowledge that might be otherwise inaccessible and undocumentable, including feelings, daydreaming, emotions, and the like, particularly as these emerge in intimate relationships. Reflexive and autoethnographic methods allow getting "into the head" and body of the researcher as a social actor, and gaining an insider's perceptive. The perspective of/from the inside is of unique significance in the research of automobility, which has notoriously centered around quantitative research of the outside of vehicles. This type of research has systematically neglected dealing with activities, emotions, and perceptions that occur inside the car (Sheller 2004 and other chapters in Featherstone, Thrift, and Urry 2005).

In de Certeau's terms, observing and documenting everyday occurrences that happen to one's self amount to a "tactic." Unlike the "strategy," which is an institutional activity that has power working for it (scientific discourse included), the tactics are everyday activities that are temporal and elusive; they are in the service of individuals who try to cope with hegemonic systems and "turn events into opportunities" (xix)

Second, because I am an independent scholar, my research expenses are funded solely by my personal budget—or, more accurately (and more relevant to this research), my family's budget. Under these limitations, researching everyday events suggests a feasible possibility and a way of making virtue out of necessity!

The Domestic(ated) Car

While little can and needs to be said here in terms of the relationships between the object of the car and the system of transportation—the former being a central actor in the latter—I wish to elaborate a bit about the relationship between the car and the family. The car offers a domestic(ated) space of intimacy that extends the family's (stationary) spaces of residence. One needs to merely observe how family members jam themselves into the car in the mornings—each on her or his way to a different destination—to see that the car is an extension of sorts of the house. Different concepts have been suggested regarding these relationships. Jean Baudrillard (1996:67), for instance, argues that "[t]he car rivals the house as an alternative zone of everyday life; the car, too, is an abode...a closed realm of intimacy" (see also Paul Virilio's notion of the "domestic car" Redhead 2004:116). For both Baudrillard and Virilio the car offers a small space that replicates the spaces (rooms) and the objects that are typical of modern houses (a "duplication of accessories" according to Virilio and Redhead [2004:116]). Likewise, the activities in the car are similar to some degree to those that transpire in everyday domestic spaces, including interactions and conversations, playing, and more. While I tend to view the car as a continuation of the family's spaces—an additional room perhaps—Baudrillard and Virilio view cars' spaces as competing or alternative spaces. The point, however, is that the interior of the car resembles the appearances and functions of the interior of house rooms, both occupied and inhabited by members of families.

In the heading above I use the infliction "domestic(ated)" simply to indicate that as a technical invention, the motorcar was not intended to perform "socially" (as is the case with many technological inventions, such as the telephone; see Fischer 1994). As Laurier and colleagues (2008:2) recently noted, cars have "become places we inhabit without necessarily being places designed to be habitable." I therefore do not argue that the car is essentially domestic, but that its appropriation by families domesticates it time and again, each and every time it is used.

Traces inside the Car

Before I attend to *inhabited motion* in and of the car, an observation regarding the domestication of the car is due. The observation concerns the inside of the parked car, which is my 1996 black Fiat Punto. It is a small car, with manual gear, and a 1.2 liter engine.

Figure 1: The outside of the Punto (Yael waving from the inside)

I bought it 6 years ago, from a sleazy wristwatch merchant, who had on his wrist a watch worth a dozen times the value of the Punto he was selling. The observation entails everyday artifacts that are not part of the car, and that their laying about therein amounts to a phase in their biography (Appadurai 1986). A list of these objects, found in my car on a morning in the spring of 2006, includes the following:

- Seven empty bottles of mineral water;
- Candy wraps, parts of dolls, and a few games on the back seat;
- Noa's new scooter, made of iron, on the back seat floor, folded in an embryonic posture;
- A number of parking tickets on the dashboard;
- A number of seminar papers in pink and yellow nylons folders below, on the floor.

The objects given above tell us something about the meaning of the interior space of the car, how is it used and by whom. If I did not indicate earlier that this list is the contents of the interior of a car, it could have easily been thought of as a list of objects found in a (messy) family storeroom. Indeed, while writing it I feel a bit embarrassed about what the readers might think of me and of the car (I want to add explanations and accounts). This tells something of the intimacy of the space, and how describing it compromises something private.

Yet the messiness of these objects actually makes some sense in terms of functionality of interior (sub)spaces. Noa's (then 6 years old) and Yael's (my younger daughter, then 2 years old) traces—including dolls, candy wraps, and Noa's new scooter—are expectedly found in the back seat, which is the children's scene in the car; parking tickets and seminar papers are in the front, near the driver, where objects related to driving and work life would be found; mineral water bottles, which indicate that I spend enough time in the car so as to have to take care of drinking water, are also located near the driver. Like my daughters' candy wraps, the bottles tell that the car is not simply a vehicle for transportation, but also a place inhabited, a place where consumptions and other practices, such as playing with dolls and talking, transpire. These subspaces suggest heterogeneity of functions and meanings in a space that is physically limited, but not dull.

On the Way to School: Interactions and Inter(e)motions

The excerpt below describes occurrences that took place during a morning like many, after I took Yael to her preschool, a short stroll from our apartment, and then returned home and took Noa in the car to school. Orly, my wife, would leave off early for work. This explains why there are only two family members partaking in the conversation reported shortly. This is a typical state of affairs of everyday life of urban families, where, during weekdays, only precious time is spent together as a whole. It is during weekends, holidays, and other special occasions that families spend time together, which indeed make for festive occasions where family-ness is celebrated (Haldrup and Larsen 2003).

The trip to Noa's school, which lasts about 15 minutes, is a routine urban drive, rather irritating due to morning congestion in the narrow and ill-maintained Jerusalem streets. The following strip includes excerpts from a conversation, and some reflection, from that period.

* * *

"Daddy?"
I hear her low voice coming from behind, though I can't see her. I've been humorously contemplating installing a double rear view mirror, like the ones Taxis have, so that I'll be able to look through two rear angles and not one, and see both the rear of the car (outside the vehicle), and the rear seat (inside the vehicle).
"Yes, sweetie," I answer.
"Can Nitzke come to visit me today?" She asks with the right touch of a melodious plea to her tone.
"Great. That's a great idea. I'll call Ruthi to see if she's available this afternoon, ok?" I reply/ask.
"Ok."
I let the car slide a bit forward toward the car in front of us. It is decorated with orange strips and black flags, and in it I see a large male driver with a large skullcap. A typical morning traffic jam, with cars honking, nerves and everything, by the old train near the Repha'im Valley junction.
I insert the cellular earphone into my ear, and I press the green bottom and hear the "peeps," which means I'm connected.
"Hi Ruthi. Good morning. What's up? It's Chaim, Noa's father."
"Hi, how've you been?" She is driving Nitzan to school and she sounds in a hurry.
"Hmm, Ruthi, I wanted to ask, hmm, Noa suggested that we meet Nitzke after school, and I wanted to ask if that's ok with you, or if you have any plans or something?"
"Hmm, I'm sorry. On Tuesdays Nizke has Judo classes."
"Aaah, ok. Simply Noa thought about it, and so I wanted to check."
"Sorry,"

"Never mind, no."
"Maybe we'll plan for another day?"
"Yes, we'll talk."
"We'll talk. Bye." I press the phone's red button.
"Mmmh, sorry sweetie," I return to Noa, and now I too add a touch of a melodious plea to my tone.
Later on our way we pass three speed bumps. They are located one after the other on the same road, a few minutes' drive away from school. Passing on top of them, preferably fast, is an amusing attraction for Noa. Like her mother, Noa loves amusement parks and rides, where she experiences tilting and jerking sensations. The bumps supply a bit of this sensation.

    As we approach the bumps I announce: "Hey, Noa, look, the bumps! Are *you* ready?!" Noa knows what's at stake, and utters a sound of excited anticipation. This is her reply. She urges me to drive faster, so passing the bumps will be felt more effectively. This I do, and the first and second bumps are a success: things inside the car—including Noa and me—are up in the air for a few milliseconds, and both the car and the things in it make the adequate noise as they land. Noa utters an excited chuckle of jubilation. The third speed bump is always a disappointment because it is rather flat. But Noa is satisfied. She had a bit of an amusement park ride experience on the way to school.

Daddy-Driver

The interaction above proceeds with an address directed at me. While Noa could have made the request directly, she begins with an address, namely "Daddy?" This is a situated choice, which, in the context of car conversation, carries particular consequences. The settings of the interior, notably the physical divide between front and back seats, implies that everyone faces the same direction and little room is available for movement. As a result, there is usually no direct eye contact between those in the back and those in the front. While the classic settings of interpersonal communication are those of face-to-face interaction (Goffman 1959), the settings in the car create a normative condition where we have face-to-back interaction (between front and back seat occupants, and side-to-side interaction between same-seat occupants). The lack of direct eye contact, so central to face-to-face interaction, means that interactants are not aware of each others' availabilities in terms of engaging in conversation. This condition requires that more checking be done before actually engaging in interaction. This is why an address is certainly in place, both checking and demanding my availability.

    But addressing me specifically as "daddy" (*aba* in Hebrew) also establishes gendered identities and social roles. The evocative "daddy" is an utterance that (re)establishes simultaneously the roles of child and male parent. These roles index the social system of which they are a part—that of the family, which is hereby being performed. The reply, "Yes, sweetie," confirms this. Father (*aba*) is now available for conversation: he confirms that there's an open channel of communication and he acknowledges occupying the role of father in the interaction.

Performed inside the car, this exchange establishes the power relations that are characteristic of both the automobility and the family as patriarchal systems. It is no coincidence that the child (in this case a female) is at the back seat, and the parent (in this case a male) is up front by the wheel. Occupants of the back seats enjoy less privileges in terms of viewing the road, and have far less access to the car's systems and devices (both those relating to driving and other features such as playing the radio and the CD).

These power relations are shaped by the "architecture of visibility" (Laurier et al. 2008:9), which form the situated politics of viewer/viewed. As I indicated earlier, face-to-face interactions are infrequent in the car, and the configuration of looks between the front and back seats usually includes mediated, face-to-mirror interaction. When in the car, my daughters routinely try to avoid being seen by me through the mirror, which they accomplish by squeezing themselves to the sides of the back seat. Also, they sometimes whisper to each other. They thus practice whatever freedom they have by avoiding my visual (and acoustic) surveillance. This is why I sometimes think of the double-lens mirror that is mentioned in the excerpt. These mirrors, which are usually installed in taxi cabs, allow a broad view of the back seat. Such optical devices indicate that the space of the vehicle's back seat is as much a sight of visibility as is the road, or, put differently, that for taxi drivers the inhabited road should be monitored inasmuch as the back seat.

Noa then proceeds with the request, which concerns arrangements for her to meet her friend Nitzan (fondly nicknamed Nitzke) later that day. The request gives us a clue as to what is on the mind of the 6-year-old passenger. On the way to school, Noa is already contemplating the way *back* from school. It might be that she is bringing together the beginning and the conclusion of her school day, which brings her to contemplate what to do in the afternoon. It might also be that she had made previous requests to see Nitzan, before we entered the car. Car conversations oftentimes reverberate conversations that had occurred earlier, both in prior trips and before embarking (Laurier et al. 2008:18). This occurs often in our family, as different moods and emotions, such as Noa's or Yael's frustration when we do not agree to something they want in the morning, or our frustration at their slow pace of getting ready to leave the house, are carried from the apartment into the car.

Having Nitzke come over has consequences in terms of both the transportation and the family systems: a positive answer confirms the parent's approval to have a friend over, and the driver's approval of picking up both girls after school, and taking the guest back to her home when the visit ends. These are of course different considerations that demand different consents, and relate to the systems evoked in and through the interaction.

Calling Nitzke's Mom

My consent is followed by an action, namely contacting Ruthi, Nitzan's mother. But before I do so, there are things that need to be attended to immediately. The jammed traffic has begun moving slowly, and I slightly lift my leg from the brake pedal and let the car slide forward a bit. Here is a case where events that concern the system of transportation, and occur outside the car/to the car, impinge on the interaction inside it, and demand the driver's attention. As Laurier (2004) observes in his research on people who spend hours doing office work in the car while driving UK highways, occasionally traffic-related occurrences intervene with the office work done in the car. In these occasions, the attention of those by the wheel shifts from office work to driving, or, in the terms employed here, from the roles that relate to the work system to the roles that relate to the transportation system. Although plain, the maneuver requires my action momentarily, and I am drawn from the inside of the car, where Noa and I are interacting, to the traffic (interacting) outside it. This is a shift between the roles of the parent and the driver, where the latter's perspective now assumes the foreground. Oftentimes these shifts are marked by such utterances as "just a sec, sweetie, I'm driving."

The driver is establishing communication with someone outside the car via the cellular telephone, but at the same time he is also looking outside. What the driver now sees is a car that has a number of political bumper stickers and ribbons and flags on it (in light of the heated culture of political bumper stickers in Israel, this observation is common, more so in Jerusalem). The small black flags and the orange-colored ribbons represent ultra-right-wing political association (usually stuck on cars of orthodox Jewish settlers in the Occupied Territories). Like political bumpers, flags and ribbons make use of the performative quality of the infrastructure of automobility, which is a consequence of its high degree of visibility. These (political) communicative devices teach us that driving is as much about seeing and showing as it is about getting from one place to another, and that much of what goes on in the road is, one way or another, political.

A feature of the dominant mode of automobility's visuality concerns car windows' two-way transparency. Not only are bumper stickers available for observation, but also some of the *inside* of others' cars. Here it gets tricky, as Katz (1999) shows, because what we see is usually ambiguous, a fact that allows speculations and projections, and contributes to the construction of other cars' interiors as fertile resources for imagination and daydreaming.

Finally, there is also a *reflexive* quality to the activity of looking *at* and *into* others' cars. For it might well be that they, too, are looking *at* and into *our* car. Here again, the (external) appearances of the car and the (internal) contents are interlinked.

I will refrain from elaborating further on the conversation with Ruthi, but I will indicate that throughout it the two roles—now (with two parents on the line) doubled—are in (inter)action. Both Ruthi and I are parents and drivers simultaneously and intermittently. We both need to take care of our families and our moving cars, and we both have to talk and coordinate activities with each other as both parents *and* drivers.

Speeding over Speed Bumps

The last event reported in the excerpt concerns the speed bumps we pass on the way to school. Here the roles of father and driver are juxtaposed in a way that I find fascinating. Speed bumps, like traffic lights, lanes and signposts, are an integral and mundane part of the transportation infrastructure. This means that we usually pass them without noticing. Yet sometimes and under particular circumstances, we "turn events into opportunities" (de Certeau 1984:xix), or in terms of objectification, we embed objects into our life worlds (Tilley 2006:60). On this morning, the particular circumstances that are involved are emotional, and concern my feelings as a father toward my daughter. Recall that a few minutes ago I was not able to successfully complete Noa's request regarding meeting her friend. And I now feel guilty of having disappointed her. It is with these feelings, *fatherly emotions*, that I approach the speed bumps as a *driver*. The emotions play a pivotal role here in shaping the driver's decisions and behaviors (Sheller 2004).

As we approach the speed bumps, I draw Noa's attention, foregrounding the bumps and the occurrence of passing over them against the routine of "everyday" car travel. This is how I succeed in awaking in Noa a sense of anticipation in the midst of a routine. The latter is fragmented *from within*, by highlighting one of its elements, and suggesting an encounter that is improvised and unexpected. A truer phenomenologist than me would point out the evocation of vertical motion in what is otherwise a plain of horizontal movements, in disrupting the order of everyday driving.

In this case the driver, through the set of possibilities that are available to him, *helps* the father (defined as well by a set of possibilities and commitments), in becoming a *satisfying/satisfied parent*. If the plan to meet Noa's friend has failed, perhaps there is something else that father can do to make her happy. The speed bumps emerge in the driver's consciousness as a timely resource, and the driver speeds the car in order to make the most of this opportunity. As vividly described in the excerpt, this is indeed what happens. Passing on the speed bumps in a speed that is higher than a "routine" speed produces the hectic consequences—psychical and embodied— that the father had wished for. And the daughter is merry.

What is so interesting here is the quality of the interconnection between the roles performed by the person behind the wheel. Unlike most instances that come to mind (including the example above and Laurier's [2004] examples), the driver here *lends help* to the father. The roles that are at play here are interestingly *complementary, or mutually enhancing*, and not exclusive or impinging.

Conclusions: Driver-Father

In this chapter I explored the domestic(ated) car as a lived and inhabited space. Following Urry (2004), I employed a systemic autoethnographic approach in the capacity of observing and understanding spaces, interactions, materials, and roles in and around the car. This approach allows a synthetic, rather than analytic view, that celebrates diversity and multiplicity. This was evinced in the twofold roles I examined: that of the parent/child, and that of the driver/passenger. These roles were both enabled by and embodied through the two systems found relevant to this inquiry: family and automobility. Of course, in different empirical explorations different systems may emerge as more or less informing, and with them different roles, meanings, and practices.

Studying the space of the car requires overcoming the externalities of cars (Miller 2001), or surfaces (Warnier 2006), and attending with detail to what is found and to what occurs inside the car. Cars have been traditionally studied from the outside, in terms of their technical performances, and their relation to transportation infrastructures. Acknowledging interior life and sociality of cars by way of autoethnography means weaving their inside and their outside together (it is in the inside of the "container" that things are transformed, mixed, and sometimes assimilated [see Warnier 2006]), and reconnecting the personal quality of the car with the public domain of the road. As Laurier and colleagues (2008:3) nicely put it, "the outside doesn't happen without the inside: without the local organization and activity of the car, the external concerns of those who study transport, disappear." This was achieved by attending to two excerpts that move from looking *at* the inside of the car, to looking *in* it and *from* it.

Exploring car travel through a reflexive ethnographic approach, which specifically captured a father-daughter conversation, turned out to be illuminating because the conversation was produced in the car. More conceptually accurate, my exchanges with Noa are illuminating not because they took place simply *in* the car, but because they are *part* of the events that take place therein, including driving and being a passenger. Thus conceptualized, the ontological status of the "car conversation" carries twofold consequences: it teaches us about the (situated) nature of conversation and about and the (interactional) nature of car travel. As I indicated in the introduction, this choice is best thought of in terms of a de Certeauean "tactic," and not as a research method; it was not conducted in order to learn about a given condition, as it was an act of endowing the condition with meaning.

Acknowledgments

I am wholeheartedly indebted to Carolin Aronis-Reinherz, Gonen Hacohen, and Orly Noy for sensitive reading and insightful comments to this chapter. All errors are my own.

References

Appadurai, Arjun. 1986. *The Social Life of Things: Commodities in Cultural Perspective.* Cambridge: Cambridge University Press.
Baudrillard, Jean. 1996. *The System of Objects,* translated by J. Benedict. London: Verso.
de Certeau, Michel. 1984. *The Practice of Everyday Life,* translated by S. Rendall. Berkeley: University of California Press.
Featherstone, Mike, Nigel Thrift, and John Urry. 2005. *Automobilities.* Thousand Oaks, CA: Sage.
Fischer, Claude. 1994. *America Calling: A Social History of the Telephone to 1940.* Berkeley: University of California Press.
Goffman, Erving. 1959. *The Presentation of Self in Everyday Life.* Garden City, NY: Doubleday.
Haldrup, Michael and Jonas Larsen. 2003. "The Family Gaze." *Tourist Studies,* 3:23–45.
Katz, Jack. 1999. *How Emotions Work.* Chicago: University of Chicago Press.
Laurier, Eric. 2004. "Doing Office Work on the Motorway." *Theory, Culture & Society,* 21:261–277.
Laurier, Eric, Hayden Lorimer, Barry Brown, Owain Jones, Oskar Juhlin, Allyson Noble, Mark Perry, Daniele Pica, Philippe Sormani, Ignaz Strebel et al. 2008. "Driving and

'Passengering': Notes on the Ordinary Organization of Car Travel." *Mobilities,* 3:1-23.
Miller, Daniel (Ed.). 2001. *Car Cultures.* Oxford: Berg.
Redhead, Steve. 2004. *The Paul Virilio Reader.* New York: Columbia University Press.
Sheller, Mimi. 2004. "Automotive Emotions: Feeling the Car." *Theory, Culture & Society,* 21:221-242.
Tilley, Chris. 2006. "Objectification." Pp. 60-73 in *Handbook of Material Culture,* edited by C. Tilley, W. Keane, S. Kuechler-Fogden, and M. Rowlands. Thousand Oaks, CA: Sage.
Urry, John. 2000. *Sociology beyond Societies: Mobilities for the Twenty-First Century.* London: Routledge.
——. 2004. "The 'System' of Automobility." *Theory, Culture & Society,* 21:25-39.
Warnier, Jean-Pierre. 2006. "Inside and Outside: Surfaces and Containers." Pp. 186-195 in *Handbook of Material Culture,* edited by Chris Tilley et al. London; Thousand Oaks, CA: Sage.

8
The Screen Deconstructed: Video-Based Studies of the Malleable Screen

Dylan Tutt and Jon Hindmarsh

In a small village in Kent a window cleaner is coming to the end of his round, completing the final cluster of houses along a winding country road as the night draws in. As he stands level with the first storey windows of a house, he reaches into his back pocket with his right hand and takes out a mobile phone, while his left grips the ladder. After a cursory glance at the time displayed on screen, he turns his phone around so that it faces outward. He then transfers it into the clutch of his supporting left hand with the phone's screen still facing away. The reason for this peculiar action becomes apparent as the light display illuminates the panes, enabling him to wipe over the windows with his free right hand to ensure a smear-free finish in the diminishing daylight.

* * *

This simple ethnographic example might serve to illustrate the social shaping of a ubiquitous electronic device, the mobile phone, in which the screen display is used to serve the everyday needs and work of a user, which perhaps deviate from the prescribed or conventional functionality of the medium (see also Tutt 2005). In a similar vein Oudshoorn and Pinch (2003) discuss the different ways an alarm clock can be put to use: to wake us up on time, to trigger a bomb, or to "signify what time it was" (Chuck D. with Jah 1997:89) in the case of a stopped alarm clock pinned to the shirt of rapper Flavor Flav. They do so to explain that "there may be one dominant use of a technology, or a prescribed one, or a use that confirms the manufacturer's warranty, but there is no one essential use that can be deduced from the artefact itself" (Oudshoorn and Pinch 2003:1–2). Indeed research into the social shaping of technology, or the coconstruction of users and technology, has long looked beyond technological determinist views of technology and essentialist views of users' identities. However, MacKenzie and Wajcman (2002:xvi) fear that this "acceptance of the overall notion of the social shaping of technology may shut off empirical inquiry into the specific ways in which this shaping takes place."

In this chapter we want to illustrate one approach to the study of the situated and contingent use of material objects and technologies: an approach that encourages a close and detailed consideration of interaction with and around artifacts. This approach is drawn from ethnomethodology and conversation analysis (see Garfinkel 1967; Sacks 1992) and makes use of the features of video[1] to capture, replay, and represent sociomaterial practices. Indeed we wish to highlight some of the value of video for exploring aspects of material culture.

Following our opening ethnographic tale, we will focus the chapter around the study of screens and, in particular, computer screens in use. The increasingly flexible and innovative designs of media devices, from mobile phones to tabletops, stretch and push the screen into different contexts of use. The growth of pervasive computing—ubiquitous, near-invisible computers embedded into the environment—is providing new forms of social and mediated interaction. In addition to a malleable and manageable boundary between the situated and virtual, the material object of the screen also provides multiple resources for communication in everyday social interaction, which is our focus in this chapter. To this end we will confront and scrutinize people's everyday methods and manipulations of the screen through close analysis of video data gathered during recent projects, which have researched media use in different workplace environments. We suggest that video provides unprecedented access to consider the situated use of the screen. Furthermore we suggest that adopting an analytic orientation drawn from ethnomethodology and conversation analysis[2] encourages researchers of sociomaterial practices to reach beyond the screen itself to consider the embodied, interactional, and real-time practices that bring the screen into our lives.

The chapter is organized around two empirical examples. The first comes from fieldwork with a video forensics analyst and it is used to illustrate the distinctive utility of video in re-presenting interaction with and around the screen. The second is taken from a video-based study of collaborative data analysis sessions among teams of social scientists. It is used to highlight the ways in which a more detailed consideration of the sequential organization of interaction (drawn from ethnomethodology and conversation analysis) can deliver a rather distinctive treatment of the screen and objects more generally.

"The Equivalent of That": Taking a Slanted View of Things

The recording of audiovisual data presents some unique advantages to researchers. As Grimshaw (1982) suggested, these kinds of data have both *density* and *permanence*; that is to say that they have the advantage of capturing multiple features of an event or scene (talk, body movement, physical context, etc.) and the advantage of being able to replay that version of the events over and over again. The value of these properties is multifaceted, but one key advantage of the permanence of the record is that it is possible to share that version of events with others, to open it up for scrutiny among a community of peers. While in textual form we cannot share the original record, in the following instance we can display a number of "images" created directly from the record. We will do this to highlight a more substantive point: that, rather counterintuitively, the content on screen at a particular moment is not necessarily key to our understanding of the screen at that particular moment. In the case at hand, the physical movement and manipulation of the screen, and the angle it can be seen, is just as critical for the ongoing discussion, and it is used by the participants to illustrate a previous state of affairs, or rather an absent quality of the video data.

Transcript 1

1	Mike:	Moss & Brown (.) they thought that their data was useless
2		(0.8) <u>because</u> (.) of the fault (0.5) a technical fault
3	Derek:	with wha[t?
4	Mike:	[wu- or: a limitation on the system they were
5		viewing it on .hhh (0.5) and that's quite common
6	Derek:	hmm
7	Mike:	erm people will (0.3) >°you know say< it jumps about it's
8		no good or hhh >you know< potentially the tape <u>might</u>
9		jump around a lot (.) but once you've processed that and
10		stabilised it you can actually get a clearer idea of
11		what's going on (.) where if you throw away all th- the
12		garbage (0.5) and leave the good bits in the middle
13	Derek:	yeah
14	Mike:	you can get an idea °of what- what's happening hhh (.)
15		so that was >that was< the original incident (.) as I say
16		it- it was virtually: **hhh if you tilt that back that-**
17		**that's pretty much what <u>they</u> could see**
18		(0.8)
19	Derek:	Right heh[e::: .hhh
20	Mike:	[the equivalent of <u>that</u>
21	Derek:	heh
22	Mike:	so that- <u>yeah</u> **it's a car park and some cars** <u>but</u>
23	Derek:	excellent
24	Mike:	°I mean (.) the give away to me was that the <u>numbers</u> were
25		dark

118   Tutt and Hindmarsh

26 Derek: ↑right
27 Mike: so (.) something's wrong there so I had a look at it

Here we see some of the uses of the screen as a material object, which can be slanted, moved up and down, and opened and closed during conversation. The example comes from our research into the different methods of video analysis practiced by video experts and practitioners from a variety of fields and professions. It involves Mike, a video forensics analyst, trying to explain video processing to a "lay" researcher, Derek, who is without technical or engineering experience. As we join the action, Mike is trying to describe the work involved in "processing" video data, damaged CCTV footage of a car park, for a particular client. The names of individuals and organizations have been altered in order to preserve their identities.

Image 1 (00:38.35)   **Image 2** (00:39.36)

(L. 16)   hhh if you tilt that back that-

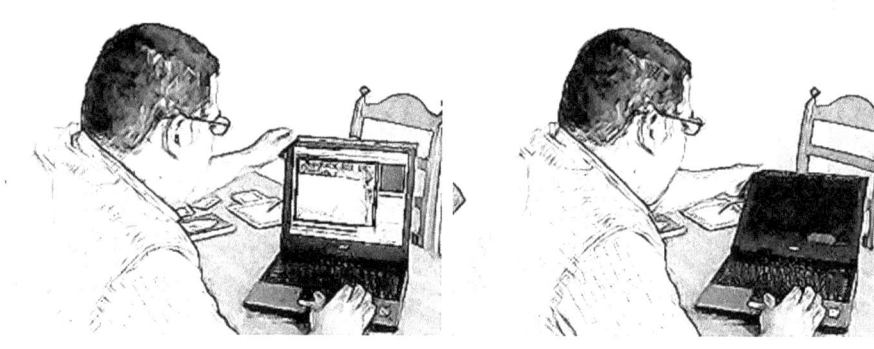

In the transcript, Mike is recounting the story of how a client came to him suspecting a problem with their security video system and needed him to "reclaim," or have processed, video data of an incident that occurred on their site car park. Having verbally described video processing (L.7-12) to Derek, Mike begins to visually communicate the difference in quality between the damaged video data and the processed video with the use of the laptop. Since Mike has the processed footage on screen, but does not have the damaged video data to show Derek, he cleverly improvises with the tools at hand. Following an audible exhalation, Mike rests his left hand on the top of the laptop screen (Image 1) and slowly pushes the screen backward, accompanied with the utterance "if you tilt that back that-" until, at an acute viewing angle, the screen image blackens (Image 2).

Mike emphasizes this covering of black across the screen, as being "pretty much what they could see" (L.17), by dragging a flat palm gesture back and forth across the screen (Images 3-4).

Image 3 (00:40.49)   **Image 4** (00:40.99)   **Image 5** (00:41.66)

(L.17) pretty much what they could see

Following this utterance Mike temporarily stops the gesture, of the flattened palm hovering over the surface of the screen, for 0.8 seconds. During this time he turns his head toward Derek (see Image 5) and smiles at him. The smile helps communicate the element of "play" in his actions, for this is a deliberate manipulation of the screen and Derek is meant to see degradation in picture clarity, which elicits laughter from Derek in response (L.19). He clarifies further, by restarting the hand gesture for one final sweep, that the damaged video looked like "the equivalent of that" (L.20).

Image 6 (00:44.59)   **Image 7** (00:45.24)   **Image 8** (00:46.29)

(L.22) yeah it's a car park and some cars but

Mike makes use of the mobility and malleability afforded by the laptop screen, to transform the picture quality and communicate to Derek "the equivalent of that" visible in the damaged video data. For the purposes of his improvised visual demonstration, therefore, the end effect—the blackened, tilted screen—stands in for the damaged video.

Mike then pulls the screen forward to the normal viewing angle to reveal again the processed video image. Derek's utterance "excellent" is actually produced before the screen is pulled upward, which nicely demonstrates his understanding of Mike's use of the screen in this sequence and of what is coming into view next. There is no great moment of "revealment" because, unlike a conjuring trick, Derek has already seen the strings and has been shown how the effect works—the contrast has been made already.

With the screen upright, Mike now describes details in the processed video. It is only now, when pointing at the screen and the number plates on the cars (L.24-5), that Mike's hand gestures are to be seen as referencing phenomena in the video *on-screen* rather than as being part of, or properties of, the *screen* itself (Image 9).

That the screen's visibility is deflected and dulled when the laptop screen is slanted at an acute angle is routinely characterized as a "problem" by designers and users alike and yet here we see how this property can be used to support the local situated demands of the conversation at hand. It is an unforeseen, unintended, or at least unusual "quality" of the screen during social interaction.

Image 9 (00:48.66)

(L. 24-5) was that the numbers were dark

The value of recording this encounter using video, then, is that we are able to present these interactional practices in a form that readily retains the essential characteristics of the original encounter and that displays how these properties of the screen are treated in practice. The occasion of use, and our rendering, is then open for scrutiny, and potentially reinterpretation, by readers. So the video provides possibilities to render visible sociomaterial practices in ways that standard field notes do not allow.

We would like to turn now to the ways in which particular approaches to the analysis of audiovisual recordings of everyday work and interaction make possible the examination of the moment-to-moment constitution of material features. Before doing that, however, we need to introduce some key features of this approach.

Studying the Screen at Work

Suchman (2007:276) traces how the turn toward the social by computer scientists in the 1980s coincided with a growing interest in the "material grounds of sociality" among social scientists, particularly by ethnomethodologists and conversation analysts who already acknowledged the importance of nonverbal action in the organization of face-to-face interaction. These interests led to the emergence of a corpus of workplace studies that attend to the accomplishment of work and interaction in various complex organizational domains, such as air traffic control rooms, emergency dispatch centers, newsrooms, and hospitals (for relevant reviews, see Hindmarsh and Heath 2007; Luff, Hindmarsh, and Heath 2000). This work has helped explicate the different ways in which objects and technologies, such as the screen, are collaboratively used to help coordinate everyday workplace practices. Taking on Hughes and colleagues' (1994:431) premise that the greatest strength of ethnography is "its ability to make visible the 'real world' sociality of a setting," workplace studies set out to study interactions and relations among work, technology, and organization. Many of these studies draw on the use of audiovisual recordings of everyday work to unpack these relations. While ethnography and video-based studies share a commitment to the study of social interaction, conventional observational fieldwork cannot provide a data source from which to sufficiently analyze detailed social interaction with and around technology. Indeed "it is not possible to recover the details of talk through field observation alone, and if it is relevant to consider how people orient bodily, point to objects, grasp artefacts, and in other ways articulate an action...it is unlikely that one could grasp little more than passing sense of what happened" (Heath and Hindmarsh 2002:102).

There are many ways in which audiovisual recordings can be analyzed. However, ethnomethodology and conversation analysis enable the analyst to take the 'density' of the data seriously and thus to consider the interrelationships between talk, gesture, movement, objects, ecology, and the like. While they do not involve a method per se, they do provide "a methodological orientation from which to view 'naturally occurring' activities and events" (Heath and Hindmarsh 2002:110).

At the cornerstone of this approach is the recognition that social interaction is *sequentially organized*. This is to say, each and every action is organized (and expected by coparticipants to be organized) in the light of immediately preceding actions and in turn provides the context in which subsequent action will emerge. Thus, as Heritage (1984) suggests, the conduct of participants in interaction is doubly contextual: both context-shaped and context-renewing. In designing contributions to interaction, then, parties display their understandings of prior action. This is also a neat methodological resource to enable researchers to take participants' orientations seriously. As Sacks, Schegloff, and Jefferson (1974:728) suggest, "while understandings of other turns' talk are displayed to co-participants, they are available as well to professional analysts, who are thereby provided a proof criterion (and a search procedure) for the analysis of what a turn's talk is occupied with."

Bodily conduct cannot be treated in isolation from talk. While such conduct may not arise "turn by turn," it can only be understood, by both participants and researchers alike, with regard to its location within the developing course of interaction. The sense of particular gestures is not entailed in the physical form of the movement, but rather derives from the way in which the action is positioned with respect to the immediately preceding and concurrent conduct (visual, vocal, or a combination of both). Similarly, reference to, and the use of, objects is produced with respect to the actions of the coparticipant, then and there, and achieves its sense and impact by virtue of its sensitivity to the local framework of activity (see Goodwin 1995; Hindmarsh and Heath 2000). This allows us to examine the ways in which individuals in interaction with others literally make sense of objects in the local milieu. So, while many approaches to the study of practice consider the organization of activities in detail, the approach that we adopt pays particular attention to the real-time organization of interactional practices. In the next fragment we illustrate the value of this approach when applied to the consideration of sociomaterial practices, and in particular the ways in which the screen is constituted by virtue of embodied conduct at the interface.

Gesture at the Interface: Invoking the World Off-Screen

To continue the focus of this chapter on the hidden uses and aspects of the screen, we will now closely analyze interactions from a collaborative data analysis session among a team of social scientists working with video data (which are referred to as "data sessions"). These data were collected as part of a project concerned with designing and developing e-social science collaboration tools to allow distributed teams of researchers to "virtually" discuss and analyze video and document data in real time (see Hindmarsh 2008). It is in this context that we have begun to consider the role of the standardized screen, in contrast to what may be offered through digital surfaces of tables (tabletop technologies such as Microsoft Surface), walls, or rooms, in the case of high-end commercial room conference systems such as Hewlett Packard's Halo or Cisco's Telepresence. While these alternatives offer ways to bridge or reformulate social interaction over distance—and are based on different surfaces and incarnations of the screen, rather than on the exclusion of the screen from these ecologies of actions—we have also been thinking about what the screen, and its physical properties, offer coparticipants as a shared object and site of orientation in social interaction. In other words, we have been concerned with how surfaces and screens coherently support communication practices for both copresent and remote working groups, attending to use of the digital surface of the screen and communication through it. During this work we have considered, for example, how local gestures and action around the screen can be critical to the ongoing distributed work of the teams during remote meetings (Tutt et al. 2007) and how producing enactments of the video data on-screen—and having this depiction of action seen as such—is an integral part of collaborative video analysis, or group data sessions (Tutt and Hindmarsh 2008). Here we consider how participants shape the ways in which the screen is seen.

This example shows how the sense and significance of the screen can be transformed, moment by moment, through the embodied actions of using it within everyday work. Here, four social scientists are analyzing video data, but they are split over two remote sites. The research team is studying video fragments of an urban car journey, and as we join the action, they are interested in the organization of directions given by one passenger. Henry (at the local site) faces the problem of getting others (at both the local and remote sites) to see something that he has noticed in the video data: evidence of "no entry" to a specific side street. Eddie (at the remote site) questions how Henry (at the local site) knows that there is "no entry" to a road in the route under examination, and Henry and Ben (both at the local site) look for evidence in the video. Elsewhere we have considered the implications of this analysis for understanding mediated communication (see Tutt et al. 2007), but here we focus on the interaction with and around the screen by Ben and Henry.

Transcript 2

```
1  Henry:   that was it there I think
2  Ben:     °oh was it
3  Henry:   yep
4           (3.1)
5  Henry:   we're just going back a bit
6           (3.8)
7  Henry:   °mark it on the screen°
8           (0.8)
9  Ben:     where ↑is it?
10 Eddie:   yeah so there's right hand turn markings on the
11          road there
12          (0.4)
13 Henry:   °back a bit°
14 Ben:     Henry thinks he can see it °but I'm no(t)°
15 Eddie:   yeah there's a (.) you- there's markings ↑on the
16          road (1.9) so you can do a right turn
17          (1.1)
18 Henry:   there
19 Ben:     have we ↑missed it have we h[ere?
20 Henry:                              [no no:: (.) that's
21          just coming up
```

In data sessions, participants are routinely called upon to ground their analytic claims in observable evidence in the video data. Therefore much of the work of data sessions involves getting others to see such evidence. This often involves the rewinding, pausing, and playing of the fragment at moments relevant to the observation being made. This is considerably complicated when the phenomena being pointed out are in the moving video rather than a paused image and thus may only be on screen for a moment or two. In this case, the matter is further complicated as the data are from a moving vehicle and they are trying to spot something at the side of the road—thus it is at a distance from the camcorder (therefore small) and only visible momentarily (as the car passes by). When Ben plays the clip, Henry leans in to the screen to prepare to spot and point out evidence of "no entry." Moments later he reaches out and points to the laptop screen. We are particularly interested here in Henry's gestures in relation to the screen (and the playing video).

As he points to the screen, Henry leans toward Ben and quietly says to him: "that was it there I think" (L.1). Ben's hand immediately moves to the playback controls to rewind the fragment, thereby displaying his understanding that the relevant moment in the video has passed. When we look a little closer at this interaction, we see that during the word "think" ("that was it there I think" [L.1]), Henry transforms his finger gesture into a route plotter. For, the pointing finger moves from temporally marking the feature as it appears on screen, to tracking it in the moving video as it passes off-screen. In this way he is indicating not just temporally but also *spatially* the content Ben has "just missed" in the video—showing where the feature would be if Ben were to imagine the passing route trailing off-screen. This fulfils a practical purpose in indicating (spatially rather than temporally) to Ben where the video needs to be rewound back to in order for the feature to arrive back into view. So Henry transforms his referential practice of pointing to the screen into a representational gesture that "marks up" a virtual route extending out of the screen.

Image 10 (C1- 00:37.90)   **Image 11** (00:41.33)   **Image 12** (00:43.57)

Image 13 a and **b** (C2- 00:52.85)   **Image 14** (00:53.84)

(L. 18) there

What we are suggesting is that hand gestures can be used to transform the ways in which someone encounters the screen and, by communicating requests for playback, enter into the practices of managing a moving screen. Consider the following sequence in which, as the clip is played back again,

Henry almost immediately points to the screen and softly suggests that Ben "mark it on the screen" using the annotation tool (Image 10). However, Ben keeps the video playing and asks quietly, "where ↑is it?" At this moment, having drawn back his pointing gesture a little (Image 11), Henry transforms the poised finger into a more substantial and cruder backward-thumbing gesture to request rewinding (Image 12). This "hitch-hiking" hand gesture suggests a sizeable rewinding required from Ben, and that the phenomenon has well past. Ben immediately pauses the clip before rewinding it.

When Ben starts to play the video fragment one more time Henry points firmly toward the screen again without producing a lexical affiliate (Images 13). Ben continues to let the clip play, and as Henry produces the deictic reference "there" (L.18), his pointing finger starts to trail to the right hand side of the moving video and curves upward (Image 14).

We can hardly guess what Henry meant by the gesture, but what we do know is how that gesture, and its relation to screen and talk, is *treated* by Ben in his next turn. Ben questions whether Henry is now indicating that the "no entry" to the road has already passed off-screen: "have we ↑missed it have we here?" (L.19). So Ben treats the gesture as revealing the movement of the feature from on to off screen. However, Henry immediately says, "no no:: (.) that's just coming up" (L.20-21), which instead suggests that Henry's pointing finger gesture is trying to project something coming up on screen, perhaps taking on the perspective of the driver and passenger, or the vehicle.

Despite the confusion, this fragment tells us something about how gesture can transform the ways in which others experience, or see, the screen. Ben can be seen as demonstrating the reflexive relationship that exists between the screen and gesture. Because of the creative way Henry construes or shapes the moving screen during the data session—using representational as well as referential gestures—his curved finger gesture toward the right of the screen is seen to reference a virtual route. In other words, the screen helps Ben form his understanding of Henry's gesture as well as vice versa. Ironically, Henry is indeed indicating a feature *off screen*, but maybe the feature is to be found on the forthcoming *yet-to-be-seen* stretch of road, rather than on the route already passed by in the video. The work of the hand in concert with the materiality of the screen—as artifact with physical shape, sides, and dimensions—is crucial in order to illustrate and imagine, through embodied practices, the virtual world *beyond* the screen.

As with the tilting by the video forensic analyst discussed earlier, here is another example of the social shaping of the screen in which the materiality or meaning of the object is subverted or mobilized. The social constitution of the screen is, of course, very different in this example, but the social interaction between the coparticipants is vital to both. This second example demonstrates how gestures around the screen transform how the screen is perceived, and how the screen is used as a resource to instruct others of how to see and operate on the display. It is intriguing that in contrast to characterizations of the screen as a window into virtual worlds, in both examples the screen is used by participants to show what is *not* there. The first example, of the tilted and "blackened" screen, is a demonstration of not being able to see the screen, and of "covering up" phenomena in the video. In the second, hand gestures are used to create and project virtual routes *off* screen from the moving video. Yet the manipulation of, or movement around, the material screen is central to transforming the ways in which participants encounter the screen. In other words, the character of the screen is transformed by virtue of the different hand gestures and their sequential placement.

Conclusion

While we say there is an outer horizon—which means other relevant objects—the relations between these objects changes with each moment of activity. Just like the physical horizon that changes with every change of perspective, the world of the natural attitude is altered by every action. (Garfinkel 1952:341)

The wide-ranging concern in sociology and cognate disciplines with the object, with new technologies, with material culture, has powerfully demonstrated the situated and socially organized character of the "stuff" of the contemporary world. That said, the ways in which objects and artifacts feature, and are treated, within sequences of interaction has been largely disregarded. The construction of the object is therefore presented independently from the very moments of social interaction in which it is embedded and experienced. As a result, the occasioned sense of objects, their fleeting and shifting character, remains unexplicated (see Hindmarsh and Heath 2000). Here we have attempted to demonstrate that the ways in which an object, such as a screen, is used, seen, and experienced is fundamentally inseparable from the actions and activities in which participants are engaged in real time. The "same" object can be described in multifarious ways, and its significance and intelligibility can shift within a single moment. Indeed, the potential properties of an object are beyond the imagination of any researcher—the possibilities are truly endless. Therefore, we suggest the need to take seriously how participants themselves treat the object at any moment, and how that treatment is relevant to the activity at hand. This provides one way of examining material culture, that is, the ways in which the social and the material come together in and through the interactional practices of work and the like.

This brief chapter has also highlighted how participants utilize the physical and malleable characteristics of the screen during social interaction—as a material object that can be touched, tilted, and trailed from by gesturing hands. Such close study of the everyday use of the screen helps take our discussions, experiences, and understandings beyond that of the conventional functionality of media and artifacts. This illustrates the value of video for those interested in studying sociomaterial practices. On the one hand video presents opportunities to replay, review, and share occasions in which objects and technologies are constituted in interaction. On the other hand video can be used to unpack the embodied practices that surround objects and technologies and that inject them with occasioned sense and significance.

Acknowledgments

This research was funded through the MiMeG ESRC e-Social Science Research Node (Award No. RES-149-25-0033). We are extremely grateful to the individuals and groups who allowed us to record their working practices. We would also like to thank Phillip Vannini for his constructive comments on an earlier version of this paper.

Notes

1. The sketch images within this chapter are based on video stills taken from our studies in different work environments. We chose to convert the images through the AKVIS sketch program to help mask the identification of research subjects while maintaining the sense of body language and gesture, and for the ease of reproduction of images in black and white.
2. The talk is transcribed according to the standard orthography used in Conversation Analysis, which was developed by Gail Jefferson (1984). It is worth noting here that the numbers in brackets represent pauses measured to the tenth of a second, a dot in brackets represents a mini-pause of less than two tenths of a second, parallel square brackets represent overlapping talk, colons represent the elongation in the production of a word, a word underlined represents louder talk, degree symbols around a series of words indicate quieter talk and the angle brackets surround talk that is spoken noticeably faster.

References

Chuck, D. with Yusuf Jah. 1997. *Fight the Power: Rap, Race and Reality.* Edinburgh: Payback Press.
Garfinkel, Harold. 1952. *The Perception of the Other: A Study in Social Order.* PhD Thesis, Harvard University, Cambridge, MA.
———. 1967. *Studies in Ethnomethodology.* Englewood Cliffs, NJ: Prentice-Hall.
Goodwin, Charles. 1995. "Seeing in Depth." *Social Studies of Science,* 25:237–274.

Grimshaw, Alan. 1982. "Sound-Image Data Records for Research on Social Interaction: Some Questions Answered." *Sociological Methods and Research,* 11:121–144.
Heath, Christian and Jon Hindmarsh. 2000. "Configuring Action in Objects: From Mutual Space to Media Space." *Mind, Culture and Activity,* 7:81–104.
———. 2002. "Analysing Interaction: Video, Ethnography and Situated Conduct." Pp. 99–121 in *Qualitative Research in Action,* edited by T. May. London: Sage.
Heath, Christian, Paul Luff, and Marcus Sanchez Svensson. 2003. "Technology and Medical Practice." *Sociology of Health and Illness,* 25:75–96.
Heritage, John. 1984. *Garfinkel and Ethnomethodology.* Cambridge: Polity Press.
Hindmarsh, Jon. 2008. "Distributed Video Analysis in Social Research." Pp. 343-361 in *The Sage Handbook of Online Research Methods,* edited by N. Fielding et al. London: Sage.
Hindmarsh, Jon and Christian Heath. 2000. "Sharing the Tools of the Trade: The Interactional Constitution of Workplace Objects." *Journal of Contemporary Ethnography,* 29:523–562.
———. 2007. "Video-Based Studies of Work Practice." *Sociology Compass,* 1:156–173.
Hughes, John, Val King, Tom Rodden, and Hans Andersen. 1994. "Moving Out from the Control Room: Ethnography in System Design." Pp. 429–439 in *Proceedings of the ACM Conference on Computer-Supported Cooperative Work (CSCW'94).* New York: ACM Press.
Jefferson, Gail. 1984. "Transcript Notation." Pp. ix-xvi in *Structures of Social Action: Studies in Conversation Analysis,* edited by J. Atkinson and J. Heritage. Cambridge: Cambridge University Press.
Luff, Paul, Jon Hindmarsh, and Christian Heath (Eds.). 2000. *Workplace Studies: Recovering Work Practice and Informing Systems Design.* Cambridge: Cambridge University Press.
MacKenzie, Donald and Judy Wajcman. 2002. "Preface to the Second Edition." Pp. 1-8 in *The Social Shaping of Technology,* edited by Donald MacKenzie and Judy Wajcman. Buckingham: Open University Press.
Oudshoorn, Nelly and Trevor Pinch. 2003. *How Users Matter: The Co-construction of Users and Technology.* London: MIT Press.
Sacks, Harvey. 1992. *Lectures in Conversation: Volumes I and II.* Oxford: Blackwell.
Sacks, Harvey, Emanuel Schegloff, and Gail Jefferson. 1974. "A Simplest Systematics for the Organization of Turn-Taking for Conversation." *Language,* 50:696–735.
Suchman, Lucy. 2007. *Human-Machine Reconfigurations: Plans and Situated Actions.* Cambridge: Cambridge University Press.
Tutt, Dylan. 2005. "Mobile Performances of a Teenager: A Study of Situated Mobile Phone Activity in the Living Room." *Convergence,* 11:58–75.
Tutt, Dylan and Jon Hindmarsh. 2008. "Reenactments at Work: Demonstrating Conduct in Data Sessions." Working paper.
Tutt, Dylan, Jon Hindmarsh, Muneeb Shaukat, and Mike Fraser. 2007. "The Distributed Work of Local Action: Interaction amongst Virtually Collocated Research Teams." Pp. 199–218 in *Proceedings of the European Conference on Computer-Supported Cooperative Work (ECSCW 2007).* Berlin: Springer.

9
Technologies of Consumption: The Social Semiotics of Turkish Shopping Malls

Tanfer Emin Tunc

About 15 years ago, Bruno Latour (1992:290–308) called on scholars to "reinstate the missing masses" of mundane artifacts, and their users, into cultural studies. Since Latour's call to action, cultural studies, as a discipline, has seen its fair share of such artifact analysis. Building on the tradition of semiological analysis as initiated by Roland Barthes, and often drawing from critical theory (structuralism first, and post-structuralism later), the cultural studies connoisseur today knows much about the ideologies inscribed onto anything from Vespa scooters to Cyborg-like bodies. Much of this literature, however, treats mundane artifacts, and consumer items in particular, as nothing but immaterial traces of broader discourses that merely interpellate meanings, thus deterministically constituting the subjectivity of their users. An approach that, instead, deeply values the materiality of objects/signs is that of social semiotics. Social semiotics differs from semiotics in several important ways (e.g., see Vannini 2007). However, the most significant of these is its treatment of signs. According to social semiotics, signs are resources for the construction of meaning. Semiotic resources are polysemic (i.e., they have multiple meanings) and polyfunctional. Therefore meanings arise in use, in interaction, and in social context. As Vannini discusses in chapter five, such a view of semiosis (i.e., meaning-making) is fundamentally technological, as it relies on the interplay between purpose-driven social agents and the material world.

One location where the material world and its social agents intersect in a frenzy of consumption is the American-style shopping mall. As Crothers (2007) has said, most people around the world will never visit the United States or meet an American in person. However, billions will consume its culture and the material symbols that comprise it through the mall, arguably the world's most heavily commercialized space (5). Despite their seemingly "mundane" uniformity, malls function as microcosms for the negotiation and performance of cultural identity through the technics (i.e., material consumer objects), techniques (i.e., tactics and strategies of consumption), and technologies (i.e., social organization and environmental application) of commerce.

As a nation that eludes easy categorization—part of Europe and Asia, but somehow not comfortable with the socially constructed label "Eurasian"; bordered by the Mediterranean Sea, but somehow not "Mediterranean";

contiguous with the Middle East, but not "Middle Eastern"—Turkey, with its ever-increasing number of shopping malls, is currently engaging in its own identity negotiation. Trapped between their Eastern heritage and the elusive/alluring "modernity" of the West, Turks have used both material symbols and consumer strategies to construct new social realities and political identities. Turkish shopping malls, in particular, have become temples of conspicuous consumption, or "spectacular societies" where the *utopia* of progress and the *dystopia* of alienation and domination have converged into a *heterotopic* countersite of postmodern iconography (Best and Kellner 2006:4–5; Foucault 1999:239). These malls "appear to be everything that they are not: [They] contrive to be public, civic places even though [they] are private and run for profit. [They] offer a place to commune and recreate, while seeking retail dollars. [In other words, they are] representations of space masquerading as representational space" (Goss 1993:40).

By presenting ethnographic data about Turkish shopping malls and interpreting findings through the analytic strategy of sociosemiotics, this study will examine the nuanced, and often obscure, interconnections that can emerge between social agents and their material worlds (Vannini 2007:121). Namely, it will explore the ways in which consumption has contributed to the production of an abstract *hyperreal* Turkish cultural identity based on illusionary landscapes and conspicuous consumption (Baudrillard 1983). In doing so, this sociosemiotic ethnographic study reveals how the transactional/transnational terrain of the mall, with its technics and techniques, has allowed Turks to attribute individual meaning to their lives and lived environments. Moreover, it elucidates how malls have facilitated the construction of a hybrid, interactive "Western" identity that has rendered consumers participants in a cycle of conspicuous consumption driven by excessive self-indulgence in the "pseudo-needs" of modernity.

Methodological Framework: The Semiotics of the Mall

I collected the ethnographic data presented in this study through fieldwork conducted in Ankara, Turkey, over the course of a year (January–December 2007). I used four major modes of data collection: observation of mall architectural structure; nonparticipant observation of mall shoppers and salesclerks (both groups serve as the basis of the vignettes included in this study); informal, unstructured, conversations with male and female mall shoppers; and interview data collected through an hour-long focus group. The focus group consisted of eight male and fifteen female students between the ages of 19 and 22, all born and raised in Turkey, and primarily from middle-class backgrounds. Students in the focus group were asked specific questions about consumption, mall versus city shopping, Turkish materialism, American cultural imperialism, and gender roles, and were allowed to speak, in Turkish, with minimal intervention from the moderator. All names used in this analysis are pseudonyms.

The data presented in this study are interpreted through the analytic strategy of social semiotics. This mode of interpretive and critical inquiry prioritizes objects in their "everyday" environments (i.e., how they are used, and how they contribute to and/or shape their social systems), and has proven in this case to be especially useful in analyzing the interaction between built environments (e.g., malls) and technology, as well as the intersection between objects and identity formation (e.g., luxury goods denoting elite social status). Sociosemiotics exposes three layers of representation—the physical, the mental, and the implied—that characterize material signs. This epistemological approach considers what objects represent physically; what they signify psychologically, culturally, and politically; and what they connote, or suggest, subliminally (Gottdiener 1995). This sociosemiotic framework is also significant because it allows for the negotiation of culture, and embraces the reality that all knowledge is contextual, reflexive, and often the result of "lived" experiences (Vannini 2007:119-121; see also chapter five this volume). In other words, consumers do not simply look at a product, or gaze at the landscape of a mall passively, accepting the meaning intended by the producer or architect. Rather, they negotiate the meanings of what they see (Blumer 1969). That is, they combine the messages embedded in signs with their own interpretive stream, or "multiple consciousness," which is a direct product of their racial, ethnic, religious, sexual, gender, socioeconomic standing, and other, equally important, markers of social situatedness (Gottdiener 1995). What emerges out of interpretative practices are contested realities and identities that make it necessary to rethink meaning and culture as a complex and contradictory set of terms operating on several simultaneous, intersecting levels: the local, the national, and the global.

Technics of the Turkish Shopping Mall

Shopping malls are perfect venues for the exploration of technics (i.e., material consumer objects) because they not only influence how social agents interact with the physical world, but also serve as loci for the ascription of meaning to the built environment. Enclosed shopping complexes are a relatively new addition to the Turkish landscape. The first mall in Turkey's capital, Ankara, was Atakule (a Seattle Space Needle, or Toronto CN Tower-type structure), which was built in 1989, followed by Karum, a more traditional shopping center, in 1991 (Erkip 2005:93). Since the mid-1990s, malls have been built in rapid succession in Turkey. Currently, about a dozen new malls are under construction in Ankara alone, and many, such as Forum, are vying to become not only the largest mall in Turkey, but also the largest mall in Europe (the title is currently held by Cevahir, which is a 4.5 million-square-foot mall located in Istanbul) (Pocock 2008). Others, such as Ankara's Armada, named after the Spanish fleet and shaped like a ship, and Istanbul's Kanyon, which resembles a canyon, have not pursued quantity of square footage, but rather quality of design. In fact, in 2004 the International Council of Shopping Centers voted Armada Europe's best shopping mall under 375,000 square feet. This award, which is prominently displayed in Armada's corporate office located in the adjacent office tower, is a constant reminder that even if "Europeanness" cannot be achieved through political negotiation, it can at least be acquired vicariously through conspicuous consumption.

Despite Turkey's low per capita income compared to North America and Europe, Turks are becoming increasingly obsessed with the "malling" phenomenon. There are two driving forces behind this trend. The first is environmental, the second social. As Erkip (2005:91) has illustrated, the "contrived spaces of the shopping mall" contrast greatly with the "incivility of the [Turkish] city street." This incivility is multilayered, and as students in the focus group noted, ranges from dangerously cracked or missing sidewalks, to death-trap crosswalks, to careless motorists, to pesky panhandlers, to muggers, to pushy pedestrians, and nonexistent parking facilities. As 19-year-old Leyla conveyed, malls offer "one-stop shopping"—all "needs" can be met in the same location. Very little walking is required in the mall; instead, technics such as elevators and escalators have replaced our feet (Gottdiener 2000:276). Thus, with its "pristine," monitored environment, the mall provides an escape from the monotony of everyday life and the troubling quotidian uncertainties of the Turkish city. For many Turks, such as 20-year-old Sibel, the mall remains a novel phenomenon that continues to transform the mundane act of shopping into a social "adventure."

Over the past two decades, Turks have sought to recreate their personal and political identities through conspicuous consumption. The mall has become a key liminal space where, on a daily basis, Turkish identity and individuality is being performed, negotiated, hybridized, and repackaged as "European" modernity: a pseudoscape where personal freedom has become equated with free enterprise and consumer choice. As Marianne Conroy (1998:63) has elucidated, "the shopping mall [has become the] premier site for the making of postmodern subjectivity—where boundaries between high and low culture are effaced, where commodities and consumer desire determine the organization of public space and the form of social exchange and, above all, where simulated experiences attenuate historical and temporal consciousness." Ismail Acar's mural at Ankara's Cepa shopping mall, entitled "Büyük Türkiye Resmi" ("Big Picture of Turkey"), is a clear representation of the postmodern subjective negotiation that is under way in Turkey (Figure 1).

Figure 1: Mural at Ankara's Cepa shopping mall

A collage of traditionalism and modernity, this whirlwind of images includes not only *historical* scenes from Turkey's past (the Byzantine, Mongolian, Selcuk, and Ottoman Empires, and World War I), but also *temporal* cultural symbols such as the evil eye amulet, whirling dervishes, and Turkish tile patterns. Shoppers with small children seem to recognize the importance of this mélange the most. As this conversation expresses, many even use the mural as an opportunity to instruct their children on Turkish culture and history:

Child: Who is that?
Mother: That is Genghis Khan, a powerful Mongolian warrior. Turks are descended from Genghis Khan.
Child: I know who that is (pointing). That is Atatürk!

Mother: Yes, who is he?
Child: He established the Republic of Turkey in 1923, and was its first president. He is the father of our country.

As 32-year-old shopper Lara noted, the most intriguing element featured in the mural are the multiple sets of seductive, female eyes: "I feel like I'm being followed. They seem to gaze at shoppers"—in this case, as they purchase goods from non-Turkish vendors such as Gloria Jean's Coffee, Bauhaus, Aldo, Calvin Klein and Lacoste in an attempt to imitate European sophistication and "modernity" through conspicuous consumption.

Despite its "Western" symbolism (i.e., skyscrapers, airplanes, and alluring, sexualized women), the mural at Cepa is meant to evoke patriotism in the face of an international capitalist takeover. At Cepa and other malls around the country, consumerism has become equated with (inter)national identity. Juxtaposed with Turkish flags and symbols are American, British, EU flags; the latter three are commonly used as a marketing ploy to sell Turkish consumers the "prestigious" technics of a modern "refined" lifestyle: bag-less vacuum cleaners, air purifiers, humidifiers, garbage compactors, and HD LCD televisions (Turkey has only recently begun to broadcast high definition programming, and the few channels that broadcast in HD can only be viewed through pricey satellite subscriptions). Although Turkish consumers realize that many of these products are unnecessary, as this exchange at Ankara's Panora mall conveys, they still feel the need to buy them, especially when an overzealous salesperson reinforces the idea that "no modern home would be complete without Gadget X":

Female Shopper: What is that?
Male Shopper: This is a waffle machine.
Female Shopper: What do you do with it? [Note: Many Turks, especially those who have never traveled abroad, do not know what waffles are since they are not part of Turkish cuisine.)
Male Shopper: You pour batter into it and make waffles. They taste like cake. I ate one when I went on that business trip to France.
Salesperson: I noticed you looking at the waffle machine. It's a neitem from France (*she indicates the French print, brand name, and EU flag on the box*). It's a very popular item. I heard that in the United States they eat waffles with syrup for breakfast! In Europe we eat them as dessert, usually with fruit or ice cream. You should buy one. They're going to be very popular this year.
Female Shopper: It looks interesting, but I don't know how to make them.
Salesperson: It's really easy. Just read the directions. Everything is explained for you.
Female Shopper: What do you think?
Male Shopper: Why not? We can pay the 150 YTL in installments [ca. US$100].
Female: Ok. Let's try it. Maybe our son will like waffles.

As this vignette illustrates, technics serve both instrumental needs (making waffles) and symbolic ones (enhancing social status and bolstering identity), and herein lies their semiotic power (see Vannini chapter five). Moreover, material technics can also represent the intersection between the technological imperative and everyday culture. The rationale in this case is that the purchase of such consumer technics/"fetish" commodities will automatically render Turkish consumers both "happy and healthy" and tastefully "in touch" with the culture of the outside world (Lefebvre 1991), in other words, ready for membership in the EU. However, what is also happening is the construction of a liminal "queer nationality" based on material goods that, in the Turkish context, have no original meaning (Appadurai 1996:169–171).

One prominent material consumer object that has played a major role in the transformation of Turkish cultural identity is the diamond solitaire engagement ring. Formerly a status symbol of the wealthy, the solitaire has now, thanks to a massive media campaign by De Beers' Diamond Trading Company (DTC), and the 2006 song "Pırlanta" ("Diamond") by Turkish pop singer Nil Karaibrahimgil, been embraced by the middle class and, as a result, can be found in every mall jewelry shop. By responding to De Beers' world-renowned sales pitch "Pırlanta Sonsuza Kadar" ("A Diamond Is Forever"), Turkish women have become active participants in the formation of an urban consumer-based identity based on material symbols. Traditionally, Turkish engagements were symbolically represented through the exchange of gold bands. However, since De Beers' marketing campaign began in the early 1990s, the solitaire has become a marital prerequisite, an expression of empowered individuality. As the students in the focus group conveyed, women of marriageable age are no longer content with the simple bands of their foremothers; 21-year-old Selda and 22-year-old Erin confirmed this phenomenon: "girls our age not only demand cars and apartments from their fiancés, but also want diamond solitaire engagement rings." As 20-year-old Melisa elucidated, this material object has become a part of the marital rite of passage: "It is part of a 'modern' woman's wardrobe. It's something that not everyone can have. It makes us feel loved." Thus, the solitaire diamond ring not only defines their status as "liberated" women, but also characterizes their class position as being "above the masses." It functions as a material symbol of what C. Wright Mills (1951:74–75) termed "status panic," "wherein members of the new middle classes come increasingly to depend on the goods they consume to express their claims to social prestige and to enforce status distinctions leveled by income."

Men of marriageable age are also experiencing social pressure to purchase solitaire rings. As 22-year-old Ali explained, "all of today's girls want diamond rings. They tell you directly and even point them out to you in jewelry shop windows. I'm a student, and I don't have any money. I have to find a job after I graduate and start saving right away. My girlfriend said that she won't marry me without a diamond ring"; 21-year-old Mustafa also expressed concern over this new material symbol: "I asked my mother why all the girls want one these days. She says it's the Western media poisoning their minds, telling them that they 'need' all these things. Love and respect have become secondary concerns. She said that when she married my father, he had nothing. They've been married for 25 years. She believes my generation is obsessed with objects. That's why many of today's marriages don't last beyond 5 years."

For many members of the Turkish middle class, purchasing a diamond ring still remains a major financial investment. Since wages are low in Turkey (in 2008, the average government employee earned around 1000 New Turkish Liras, or US$650, a month), and the cost of living in Ankara and Istanbul is quite high (comparable to the United States or Canada), under normal circumstances, the average middle-class full-time employee would not be able to afford a diamond solitaire ring. However, an artificial mechanism designed by vendors and credit card companies to "democratize consumption"—the *Taksit* system—has distracted Turkish consumers from these socioeconomic incongruencies by allowing them to indulge in material possessions previously limited to the wealthy. This payment installation plan (which can last anywhere from 6 to 24 months) facilitates both the acquisition of goods (sometimes luxury items, but mostly basic necessities like food and clothing), as well as the impersonation of class positions beyond actual income levels (i.e., social mobility, at least materialistically). Despite the obvious culture of consumerism created by this system (prices are kept high because, as customers are reminded, "you can always pay in installments"), very few Turks complain. On the contrary, they perceive it to be a beneficial extension of the private enterprise system that facilitates the acquisition of previously inaccessible goods.

Techniques of the Turkish Shopping Mall

As Miller (1998) has illustrated, shopping malls are the perfect microcosm for the examination of the techniques (i.e., tactics and strategies) of the consumer cycle (i.e., production, distribution, promotion, and retailing). Turkish food courts in particular provide an interesting case study of the intersection between identity formation and consumer technics, techniques, and technologies because they serve as stages upon which the struggles between nationalism and internationalism are enacted. The food courts found in Turkey serve as stages upon which the struggles between nationalism and internationalism are enacted. In order to be "all things to all people," Turkish malls attempt to simulate a whole range of "international" culinary experiences that will appear nonthreatening to their desired middle-class and foreign clientele (Sterne 1997:27). In addition to the expected Turkish-themed establishments, almost all food courts include a Burger King, McDonald's, Kentucky Fried Chicken, a generic Mexican restaurant, a Dominos or Pizza Hut, a Starbucks or Dunkin Donuts, a baked potato stand, a deli-type sandwich vendor, and a salad bar. While, on the surface, the Turkish food court appears to be a culinary version of the United Nations, a quick glance at the menus offered by the American establishments reveals a very different picture. Attempts to cater to the local market, such as McDonald's "McTurco," a sandwich consisting of *döner kebap* served in pita bread, and Domino's *döner kebap* topped pizza, function as contrived marketing ploys designed to tempt Turkish customers into becoming fast food addicts of sorts. This is causing two major problems in Turkey: the sudden appearance of childhood, adolescent, and adult obesity (which did not exist prior to the fast food invasion) and a clash of cultures. As observed in this vignette that transpired at Ankara's Armada Mall, children are enticed by the cheap made-in-China toys that come with "Happy Meals," and start screaming for McDonald's as soon as they smell the odor of its deep-fried food, and see its iconic golden arches:

> Child: I want to eat McDonald's! I want that toy!
> Mother: Wouldn't it be nicer if we ate here? You can eat some chicken and salad. You like chicken. I can also get you some ayran [a Turkish yogurt drink].
> Child: No. I want McDonald's. I don't want chicken. Let's go now! [The screaming continues].
> Mother: OK, OK! Come on.

Although they are aware of the fat and empty calories of fast food, Turkish parents are often helpless and usually succumb to their crying children's demands. Prior to the invasion of foreign vendors, parents dictated their children's consumption habits. This no longer seems to be the case in Turkey, especially among middle- and upper-class families with disposable income. Today, cheap plastic toys (i.e., promotional technics/tactics used by McDonald's and other fast food establishments) seem to be shaping what children, and consequently parents, consume in the retail sphere. The magnetic lure of the golden arches has rendered many Turks partly unwilling participants in the McWorld phenomenon and the technologies that sustain it.

In this context, the mall functions as a semiotic terrain that is constantly defining and redefining needs in an attempt to perpetuate the consumer cycle.

Because "foreign" fast food is usually double, and sometimes even triple, the cost of the local cuisine, eating KFC, or drinking a cappuccino from Starbucks (which costs just as much as a complete Turkish fast food meal), is not only part of the "modern" shopping and entertainment experience, but also a technique, or strategy, used by many Turks to distinguish class status through consumption. Strolling through the mall with a Frappuccino in hand has become just as important as making purchases from Laura Ashley, Marks and Spencers, or Sephora, because, as Chase (1991) has delineated, the symbolic value ascribed to prestige products often exceeds their material value: "the act of acquiring the product and the associations of the product's advertising and marketing [have] become...more [significant] than the product itself" (211–212). Products emblazoned with English words have become arguably the most potent consumer fetishes in Turkey because they "signal modernism and internationalism...the *denotative* meaning attached to the words is often secondary. What is more important is an appreciation of the language's implicit, symbolic, [*connotative*] meaning" (Alden et al. 1999:77, emphasis added). In this context, consumer identity becomes an exercise in instant semiotic recognition: a Turkish shopper sporting clothing purchased from a high-end, foreign retailer (usually emblazoned with a distinctive monogram) is automatically assumed by onlookers to possess an "elite" socioeconomic position.

As Jon Goss (1993) contends, "nature" is also semiotically important in the built retail setting because it signifies the "final frontier," a locale that preserves flora and fauna that, under normal circumstances, would not be able to survive without artificial enclosure. Foliage (palm trees, shrubs, topiaries, and flowers), water (fountains, streams, and ponds), and specialized lighting (fiber optics, chandeliers, and skylights) commonly provide a mood of civilized adventure, adding excitement to the mechanized activity of shopping (36). In the Turkish-built retail environment, architects and interior designers frequently manipulate "nature" and employ contrived exotic themes (such as the "nautical traveler" style) in an attempt to signify Western modernity and sophistication. Turkish malls can be shaped like boats (e.g., Armada), include maritime symbols (such as anchors, compasses, clouds, and sails), or be themed around escapist locales (e.g., Istanbul's Kanyon, with its sloping "canyon-esque" walls, or Ankara's Mina Sera, with its Italian Palazzo design). Thus in Turkish malls, recreating the "great natural outdoors" is a *multipurpose* technique designed to attract shoppers, exoticize the mundane task of consumption, and redefine consumer identity.

Ankara's Panora, whose slogan is "Alişverişin Doğası" (The Nature of Shopping), is probably the most blatant example of this symbolic strategy. Not only is the mall surrounded by lush gardens (which in Ankara's dry climate is truly a luxury), but it also houses two aquariums, complete with baby sharks. Moreover, Panora features a gigantic floor mosaic of sixteenth-century Ottoman admiral and cartographer Piri Reis' famous premodern world map, complete with wooden sailboats and an overhead illuminated glass dome, which, from the mall's exterior looks like Epcot Center's "Golf Ball" structure. As a female shopper in her sixties expressed, "When I come to Panora, I don't feel like I'm in Turkey. I feel like I'm in some other magical place." The use of such a theme not only suggests the possibility of traveling abroad and broadening one's horizons, at least vicariously (most Turks never leave Turkey because they cannot afford it; a 5-year passport, for example, costs US$400, which is, based on per capita earnings, the most expensive passport in the world), but also suggests that those who consume the "lifestyle" signified by mall products will become instant members of the global network of modernity. The message is clear: if Turks cannot assert themselves in the international arena politically, at least they can assert themselves materially by purchasing symbolic representations of worldly cosmopolitan sophistication, and conforming to an idealized/abstract "reality." Thus in the mall setting, "artificial" nature functions as an appropriate strategy for the construction of an elusive consumer-based identity.

Conclusion

As this ethnographic study has illustrated, consumption "serves not only to communicate one's self image to others but also to reinforce it to oneself" (Brekke and Howarth 2000:497). For fleeting moments, consumers can construct substitute identities, acting out hidden parts of their egos by purchasing symbols representing their secret hopes and subconscious desires (Chase 1991:212). Standing between the ephemeral and concrete, the mundane and exotic, and the local and global, the mall has become the perfect space to perform cultural identity, and construct personal representations, specifically because it is a place of "liminality...a state between social stations, a transitional moment in which established rules and norms are temporarily suspended" (Goss 1993:27). In Turkey, the mall is both a semiotic terrain used by retailers to market their goods and a cultural vehicle through which modernity can be expressed through consumption. If, as Jon Goss maintains, "you are what you buy," then Turks have become willing participants in this semiotic spectacle, oftentimes undermining their own traditions and histories through the consumption of vacuous status symbols and hyperreal synthetic culture (20).

While the mall has created an illusion of social cohesion based on "democratized spending," it has not contributed to the expression of personal identity based on democratic principles. It has replaced traditional forms of Turkish consumption, such as shopping in outdoor markets and on "Main Street," with an elusive landscape that has divided the population according to purchasing power. For many Turks, modernity, sophistication, and "Europeanness" have become inextricably linked to their ability to acquire status symbols—a game that continues to be reserved for the middle and upper classes. In the Turkish-built retail environment, human interaction and social equality have been transformed into the rush to consume strategically mediated technics—a perpetual cycle that is slowly being unraveled by critical and interpretive sociosemiotic analyses such as the present one.

References

Alden, Dana, L. Steenkamp, E. Jan-Benedict, and Batra Rajeev. 1999. "Brand Positioning Through Advertising in Asia, North America, and Europe: The Role of Global Consumer Culture." *Journal of Marketing,* 63:75-87.

Appadurai, Arjun. 1996. *Modernity at Large: Cultural Dimensions of Globalization.* Minneapolis: University of Minnesota Press.

Baudrillard, Jean. 1983. *Simulations.* New York: Semiotext(e).

Best, Steven and Douglas Kellner. 2006. "Debord and the Postmodern Turn: New Stages of the Spectacle." Accessed April 20, 2008:
http://www.gseis.ucla.edu/faculty/kellner/essays/debordpostmodernturn.pdf

Blumer, Herbert. 1969. *Symbolic Interactionism: Perspective and Method.* Berkeley: University of California Press.

Brekke, Kjell Arne and Richard Howarth. 2000. "The Social Contingency of Wants." *Land Economics,* 76:493-503.

Chase, John. 1991. "The Role of Consumerism in American Architecture." *Journal of Architectural Education,* 44:211-224.

Conroy, Marianne. 1998. "Discount Dreams: Factory Outlet Malls, Consumption, and the Performance of Middle-Class Identity." *Social Text,* 54:63-83.

Crothers, Lane. 2007. *Globalization and American Popular Culture.* Lanham, MD: Rowman and Littlefield.

Erkip, Feyzan. 2005. "The Rise of the Shopping Mall in Turkey." *Cities,* 22:89-108.

Foucault, Michel. 1999. "Of Other Spaces." Pp. 229-236 in *The Visual Culture Reader,* edited by Nicholas Mirzoeff. London: Routledge.

Goss, Jon. 1993. "The 'Magic of the Mall': An Analysis of Form, Function, and Meaning in the Contemporary Retail Built Environment." *Annals of the Association of American Geographers,* 83:18-47.

Gottdiener, Mark. 1995. *Postmodern Semiotics: Material Culture and the Forms of Postmodern Life.* London: Blackwell.

———. 2000. *New Forms of Consumption: Consumers, Culture, and Commodification.*

Lanham, MD: Rowman and Littlefield.
Latour, Bruno. 1992. "Where Are the Missing Masses? The Sociology of a Few Mundane Artifacts." Pp. 225-258 in *Shaping Technology/Building Society: Studies in Sociotechnical Change*, edited by Wiebe Bijker. Cambridge, MA: MIT Press.
Lefebvre, Henri. 1991. *The Social Production of Space*. London: Blackwell.
Miller, Daniel (Ed.). 1998. *Shopping, Place and Identity*. London: Routledge.
Mills, C. Wright. 1951. *White Collar: The American Middle Classes*. New York: Oxford University Press.
Pocock, Emil. 2008. "American Studies at Eastern Connecticut State University, Shopping Mall and Shopping Center Studies, World's Largest Shopping Malls." Accessed April 20, 2008. Available at: http://www.easternct.edu/depts/amerst/MallsWorld.htm
Sterne, Jonathan. 1997. "Sounds like the Mall of America: Programmed Music and the Architectonics of Commercial Space." *Ethnomusicology*, 41:22-50.
Vannini, Phillip. 2007. "Social Semiotics and Fieldwork: Method and Analytics." *Qualitative Inquiry*, 13:113-140.

10
Cultural Phenomenology and the Material Culture of Mobile Media

Ingrid Richardson and Amanda Third

Cultural studies takes as its central concern the analysis of the everyday cultural interactions of postindustrial societies. Cultural studies scholarship has given rise to a body of theoretically informed empirical research that draws upon the frameworks of, to name a few, "ethnography, anthropology, sociology, literature, feminism, Marxism, history, film criticism, psychoanalysis and semiotics" (Freccero 1999:14). As Sarah Pink (2007:18) explains—citing Pertti Alasuutari—cultural studies combines methodologies rather than appropriating a single method and "has often been described by the concept of *bricolage*; one is pragmatic and strategic in choosing and applying different methods and practices" (see also Alasuutari 1995:2, original emphasis). Historically, cultural studies scholarship has been fundamentally concerned with the ways that material objects are brought to life within the field of quotidian experience, that is, with how objects are integrated into both symbolic and material practices and take on particular historically and culturally specific meanings. Ethnographic methodologies have long constituted a mainstay of cultural studies research because they facilitate the documentation of the practices of everyday life and the experiences of "real" people in ways that dovetail with the qualitative research emphasis of cultural studies. More recently, however, in response to an empirical turn within the discipline, ethnographic methodologies have become even more fundamental to cultural studies research. Among these methodologies is the one we describe here in its application to the study of portable communication technologies: cultural phenomenology.

In its phenomenological focus, our approach is framed by the premise that every human-technology relation is a body-technology relation, invoking certain kinds of being-in-the-world, as well as ways of knowing and making that world. In particular, the work of Maurice Merleau-Ponty (1962, 1968), Don Ihde (1993), and later feminist accounts of "intercorporeality" by Gail Weiss (1999) provide effective tools for interpreting the somatic intimacy of wearable and handheld media, and the collective or sedimented mobile-user habits of a culture. These theorists consider embodiment to be under continuous modification by artifacts, tools, techniques, and more complex technological ensembles. In particular, Merleau-Ponty (1964:5) famously claimed that the body "applies itself to space like a hand to an instrument," an "application" that depends as much on the specificities of perception and bodily movement as it does on the materiality of the tool in use. In his well-known description of the blind man and his stick, Merleau-Ponty describes how the corporeal schema of the body "dilates" and "retracts" to accommodate tools:

> The blind man's stick has ceased to be an object for him and is no longer perceived for itself; its point has become an area of sensitivity, extending the scope and active radius of touch and providing a parallel to sight. In the exploration of things, the length of the stick does not enter expressively as a middle term: the blind man is aware of it through the position of objects rather than of the position of objects through it. The position of things is immediately given through the extent of the reach that carries him to it, which comprises, besides the arm's reach, the stick's range of action. (22)

This quote describes the reality of what Merleau-Ponty refers to as our corporeal or body *schema*, which is not determined by the boundaries of the material body but rather reflects the way that our corporeality changes its very reach and shape in its dynamic apprehension of tools and things in the world. Merleau-Ponty argued that this schematic is inherently open, allowing us to incorporate technologies and equipment into our own perceptual and corporeal organization. In these terms, we can see how the material and perceptual specificity of media interfaces and apparatuses are deeply integral to our individual and collectively realized corporeal schemas. In the context of a study of mobile media, it is worth noting that phenomenology conceives of movement, mobility, motility, and gesture as fundamental to our somatic involvement with the world, and integral to visual perception. Moreover, the corporeal schema is in an emergent and dynamic *relation* with our environment—it is always a mode of doing, and thus always situated and contextual.

Studying Portable Media

Portable media devices and wearable communication technologies are becoming progressively more ubiquitous and personalized, penetrating and transforming everyday cultural practices and spaces, and further disrupting distinctions between private and public, face-to-face and telepresent interaction, and actual and virtual environments. Such devices range from the standard mobile phone to highly sophisticated multimedia hybrids, personal digital assistants (PDAs), MP3 players, personal media centers, and handheld networkable game consoles. In this chapter we explore the materiality and phenomenology of mobile communication and media use, and the embodiment of handheld small screens, by considering how various modalities of use afford a range of different attitudes, postures, motilities, and body-space relations. To make these arguments, we analyze the results of an ethnographic study of Australian young people's use of the mobile phone as a technology of convergence, deploying the methodological frameworks and insights of cultural phenomenology and visual ethnography.

This study sought to document and evaluate how young Australian people use their mobile phones in the contexts of their everyday lives, along with the range of attitudes they hold toward mobile phone technologies. We undertook half-hour-long semistructured interviews with twenty-five project participants aged 18–24, asking each of them to speak freely about how, when, and where they used their mobile phone and to show us examples of their usage. In particular, we were interested in establishing how the mobile phone as a material artifact is implicated in shaping young people's routine practices. Our line of questioning, then, was motivated by a concern to delineate the various material and corporeal effects that emerge from mobile media use, and thus more broadly, the spatial, perceptual, and ontological effects of mobile media. Such questions highlight a field of enquiry that is focused on the phenomenology and medium specificity of mobile media and what could be termed their technocorporeal or *technosomatic* attributes, that is, the particular relation between mobile technology and one's embodied perception and situatedness in the world.

The study at the outset indicated that the increasing proliferation of mobile handsets and the way they prioritize specific functionalities are paralleled by highly idiosyncratic usage patterns. Young people incorporate mobile handsets into their everyday lives in ways that are intensely individualized. In this sense, mobile phones, as material objects, work in consonance with postmodern trends toward individualization and the fragmentation of media markets. We begin by first outlining the combined research strategies of cultural studies, visual ethnography, and phenomenology, the way they come together in *cultural phenomenology*, and their relevance to the theorization of mobile media and their embeddedness in material culture.

Visual ethnography defines a particular ethnographic approach that accounts for the significance of images and visual media in contemporary culture, and is linked to cultural studies in the way that the latter ascribes importance to both ethnography and the visual (and visual media specifically) in the production of meaning. The principles of visual ethnography can be flexibly brought to bear on the multisensory nature of experience and meaning, and point to the way ethnographic methods attend to the complex intermeshing of material culture, corporeality, and social relationships (Pink 2007). In this way visual ethnography provides an important connection between the ethnographic turn in visual and cultural studies and the phenomenological method. In fact, visual ethnography and cultural phenomenology share a common goal in wanting to critically account for the perceptual and sensory dimensions of everyday material culture. In our case study, then, participant interviews focused not only on the communicative or exclusively visual experience of mobile media, but also on the haptic, gestural, and embodied appropriation of the device.

Cultural phenomenology resituates embodiment and materiality within sociocultural contexts, combining the phenomenological focus on corporeality, perception, and modes of being-in-the-world with a constructivist, representational, or semiotic analysis. In Csordas's (1999:143) terms, cultural phenomenology effectively synthesizes "the immediacy of embodied experience with the multiplicity of cultural meaning in which we are always and inevitably immersed." This kind of analysis, which embeds cultural, historical, and gender specificity into our relational ontologies, has been the particular focus of corporeal feminists such as Gail Weiss (1999) and Moira Gatens (1996) and also resonates with the work of contemporary theorists of technology such as Donna Haraway (1991) (in her claim that agency is irreducibly material-semiotic) and Don Ihde (1993) (in his development of the postphenomenological method). In its critical apprehension of embodiment, cultural meaning, and the significance of our being with things and objects in the everyday, cultural phenomenology brings together the concerns of material culture, cultural studies, visual or multisensory ethnography, and phenomenology. In this chapter we suggest that such an approach is particularly salient to the exploration of our communicative and corporeal appropriation of mobile media.

A Cultural Phenomenology of Mobile Media

In the contemporary context, all new kinds of computing depend on the phenomenon of human motility and mobility, such that we ourselves become their "intimate mobile hosts" (Robertson 2005). With sedimented use of a particular mobile phone, "through an ongoing adjustment of motility," we take "the motor space of our interaction" *into* our hands and the space of our bodies, such that we come to know its model-specific characteristics in the same way that we know the placement of our own limbs and fingers (ibid.). Roberston argues that just as Merleau-Ponty maintained that "learning to type quite literally incorporates the space of the keyboard into bodily space," the space of "our mobile phones is incorporated into bodily space in the same way" (ibid.), thus effectively rearranging our corporeal schematic.

A number of new media critics have claimed that virtual communication engenders the experience of disembodiment: an experience of leaving the body behind. Similar to Roustan's experiences (this volume), and in contrast to these claims, our informants often talked about the ways their engagements with their mobile phones were shaped by the need to take account of the physical body whether, for example, to reduce factors of bodily discomfort while speaking or texting, or to position the body and/or the phone to see or hear better. Participants also talked very specifically about the "emplacement" of the mobile upon or near their body ("I always keep it in my back left pocket"), and many noted that when in silent or vibrate mode the mobile needed to be in particular visual, tactile, or aural proximity to the body. These small but constant accommodations and choreographies of the body—often experienced as interrupting the seamlessness of our mobile phone practices (having to "change ears")—throw the presence of the body into sharp relief and highlight the centrality of the body-tool relation in processes of telepresent communication. Against the notion that one can be even intermittently "without a body" when using the Internet, the telephone, or watching TV, cultural phenomenological readings of telepresent media draw attention to the ways they become part of the corporeal schematic, working on and modifying the body, and affording a qualitatively distinct sensorial engagement with the world. Quite literally, we incorporate such technologies into our everyday and mundane experience of having a body. This was highlighted by our participants' frequent reference to their mobile phone as a body part; for example, the anxiety caused by losing one's phone was often described in terms of corporeal distress or dismemberment ("it's like having my arm cut off").

Our corporeal intimacy with the handset or portable console renders it an object of tactile and kinaesthetic familiarity. Nonetheless, the growing complexity of mobile devices can also bewilder the nonexpert user. In this context the technical and ergonomic configuration of the mobile media device is significant. In the optimal embodiment relation, the device should become transparent; the best usability is one that recedes from the user's awareness, such that the liminal gap between hand and instrument goes all but unnoticed. The contrivances of the body are quite literally *built into* the blueprints and specifications of any technical device or assemblage (the arrangement of keys, hands-free usability), just as the body is manoeuvred and disciplined in a Foucauldian sense by the procedures of the apparatus (typing with the thumb, or talking via a wireless Bluetooth headset while on the move). Of phenomenological interest here is the extent to which ergonomic and stylistic differences in the design of handsets and mobile devices—the material contours of the mobile medium itself—impact upon the body-tool relation, and thus upon our apprehension and experience of (tele)presence, (techno)space, intercorporeality, and sociotechnical agency.

Interestingly, none of our participants admitted to reading their mobile phone manuals, and thus became dexterous only with those functions that were intuitive or easily incorporated into their existing corporeal competencies. This was reflected in their loyalty to the particular model or brand of handset. When upgrading their phone or replacing a lost or stolen phone, participants most often opted for the same make or model, because they were comfortable with the interface, style (fliptop, candybar, slide), menu structures, keypad, and material contours of the phone. This familiarity with the mobile device means it recedes from awareness and becomes literally and materially an incorporative aspect of the body—in Heideggerian (1977) terms, the mode of one's being-with-equipment becomes *ready-to-hand*. Whenever, and often due to the restrictions of the service provider, they were forced to acquire a different make and model, our informants spoke of the frustration, discomfort, and sensory alienation caused by the demands of an unfamiliar corporeal regime. One participant voiced her frustration quite literally in terms of the time it took her to become sensually attached to the device, saying "it took a while for me to fall in love with this phone." We can see here how the material dimensions of usability and ergonomics, and their embedment in our corporeal schematics, come together in a bodily and emotional attachment to the mobile phone. We turn now to the place of mobile screens in shaping a particular mode of haptic engagement with mobile media technologies.

Considering the number of hours that many people spend engaging with media in contemporary life, the body-screen relation may be one of the most significant relations to structure everyday practices. Televisual screens are typically *paradigmatic* of contemporary perception. As Lev Manovich (2001:20) points out, despite numerous innovations in televisual media, the window remains the archetypal interface: "Dynamic, real-time, interactive, a screen is still a screen... [A]s was the case centuries ago, we are still looking at a flat, rectangular surface, existing in the space of our body and acting as a window into another space. We still have not left the era of the screen." In phenomenological terms, we can also define some general features of what Introna and Ilharco (2004) call screen-ness. Screens are frequently a focus of our attention and concern: they literally display that which is considered relevant or worthy of notice (see Tutt and Hindmarsh this volume). This property of relevance indicates "a particular involvement in-the-world in which we dwell and within which screens come to be screens. It is not up to anyone of us to decide on the already presumed relevance of screens; that is what a screen is—a framing of relevance, a call for attention" (227). However, what the arguments of both Manovich and Introna and Ilharco do not account for is the specificity of *mobile* screens in contemporary culture. While it might seem that all contemporary screen technologies are representative of the dominant televisual and ocularcentric paradigm, we argue that mobile media screens require a corporeal schematic quite at odds with our usual habits pertaining to screens.

In the first instance, the very mobility of mobile screens demands a distinct corporeal and perceptual orientation from users. Introna and Ilharco suggest that screens of all kinds enter our involvement-in-the-world at the moment we turn them on, at which point we "sit down, quit—physically or cognitively—other activities we may have been performing, and watch the screen" (Introna and Ilharco 2004:225). In making this claim, Introna and Ilharco elide the distractedness that often shapes our interactions with screens, particular those residing in domestic or busy public spaces. However, while these kinds of screens do not necessarily demand our full attention, what we can say is that cinema, television, and the computer discipline the body into a more or less *stationary* face-to-face interaction. The generally immobile screens of desktop computers, lounge-room televisions, and home or public cinemas become a focal point of our attention and confine our bodily movement.

The mobile media screen radically skews the stationary and frontal relationship that is typical of our engagement with most screens and is marked by a minimalist mode of attention. Indeed, our turning toward them is usually momentary (reading a text or checking for a missed call) or at most requires an erratic interactivity (texting, updating a contact, selecting an MP3 track). In this respect, we could say that the mobile phone furthers the fragmentation of the frontal orientation toward media screens that is typified by the often inconstant nature of domestic television viewing. That is, whereas television consumption is structured to a certain degree by the glance (rather than the gaze), our study indicated that mobile phone usage is even more thoroughly overdetermined by this regime of viewing. The eyes of the mobile media device are more distracted by the surrounding clamor and moving objects on the street or sidewalk, by the latent, lateral, but ever-ready possibilities of incoming messages, and by the mobility of one's own body. Even the seemingly committed practice of game-play on the mobile phone is characterized by interruption and sporadic or split attention in the midst of other activities: a behavior quite distinct even from handheld console game-play on the Nintendo DualScreen (DS) or PlayStation Portable (PSP). This is recognized by the growing mobile phone game industry and its labeling of a key market as casual gamers who play at most for five minutes at a time and at irregular intervals (Hume 2005). Mobile phone use is thus rarely a dedicated practice—it is always already surrounded by other objects and activities within the spatial topography of the built environment (for a related finding in the context of gaming see Roustan this volume).

Further, despite being avid participants in online culture more generally, our informants expressed a marked reluctance to use the Web microbrowsers on their mobile phones. In part this was in response to the perception that using their mobile phones to access the Internet would incur excessive costs. However, our participants also described their disinclination in terms of certain material limitations of mobile handsets. In particular they identified the small size of the mobile screen, as well as the fact that they rarely sat and gave their mobile screen the kind of focus that best facilitates effective Internet engagement, underlining the fact that, for our young Australian informants at least, the mobile phone is conceived primarily as a technology for mobile social networking rather than online networking and "small media" consumption.

The intensely glance-based regime of viewing that has become characteristic of mobile phone engagement is also consonant with the primacy of SMS (Short Message Service) communication reported by our sample, and goes some way toward explaining the prevalence of texting among Australian users. The overwhelming majority of our participants claimed they used their phones primarily for SMS communication, well above any other feature, with voice calls falling well below utilities such as the camera and MP3 player. Given the text-based nature of SMS, this indicates that our informants conceive of and relate to their mobile phones primarily as visual, rather than aural, technologies. Our participants cited several reasons for their strong preference for SMS communication, some of which are highly specific to the Australian context. Cost was a major incentive to using text-based messaging. In Australia, voice calling rates have traditionally been quite expensive when compared to international cost structures and therefore prohibitive for young people with a limited income, whereas SMS messaging is much more affordable. However, our participants also noted that SMS communication afforded them particular modes of mobility and the capacity to use their time more efficiently, more so than voice calling, video calling, or using the Web interface of their phones to chat via Web 2.0 social networking platforms. Our sample consistently described their SMS communication practices as "conversations" consisting of between five and forty messages, with the conversational exchange unfolding over an extended period of time—sometimes over the course of an hour or so, but more frequently stretching across days or even weeks. These conversations took place on-the-go and were squeezed into brief moments seized opportunistically: waiting at a red light, while walking from one class to the next, in a brief moment of inattentiveness at a lecture, while chatting with friends face to face, and so on. Many identified the ability to take in a text message at a glance and respond at their leisure as key to their preference for mobile SMS communication.

Digital video and photograph capability via the mobile phone screen is also demanding a perceptual, visual, and cultural literacy particular to the medium. As Gerard Goggin notes, camera phones have brought about a new way of recording social and cultural contexts; they occupy "a dynamic and contingent niche in a rapidly changing scene of digital photography, image circulation, and visual culture" (Goggin 2006:153). More so than the digital camera, the mobile phone is customarily accepted as an ever-present appendage, while its multifunctionality also renders its status as camera ambiguous. The phone camera's ubiquitous visual access effectively heightens our visual awareness of the everyday, converting every situation into a potential photo or mini video pictorial storytelling opportunity (Okabe and Ito 2005). In discussing their use of the camera phone, a number of our informants claimed that although they often kept photos in their phone's memory, they never or very rarely sent MMS messages (i.e., photos or video) to friends. Rather, they preferred to show their photos to each other when face to face, or keep them as private visual memories, thus using their phone as a portable, contained, always-accessible and personalized image-archive carried on the body.

These examples of mobile screen use show that the emergent body-tool relation we have with mobile screens has seen a number of adjustments to our corporeal schematic in both communicative and motile contexts. The various postures and habitual skills surrounding mobile phone use, such as becoming adept at texting while walking, mobile game-play on-the-go, or the practice of sharing one's small screen with others, clearly undermine the more frontal and gravitational ontologies demanded by traditional screens, and also point toward the haptic qualities of mobile phone usage.

While mobile screens elicit the modification of traditional modes of screen viewing, our participants indicated that interactions with mobile phones are nonetheless structured by a *regime of visibility* that entails not just seeing with the eyes but with the whole body. Our negotiation of the mobile handset as material artifact is structured by vision, whether we are looking directly at the device or not. In phenomenological terms, we are always fully sensual and as such our ocular vision is mobilized in relation to our whole range of perceptual engagement with the world: vision exceeds the capacities of the eyes. Indeed, for phenomenology, *all* vision is tactile vision; as Merleau-Ponty

(1964) would remind us, there is no *essential* or ontic difference between vision and tele(-)vision, between touching and seeing; looking, tasting, smelling, and hearing are all variants of "handling" the world. Indeed, palpating with the eyes—*a prehension, a prise*—literally translates as *holding* (133).

In this context, a notable aspect of the embodiment of mobile phones and screens is the manner in which what is seen on the screen is tangibly and contingently dependent on the hand's movement and dexterity. There is no doubt that phenomenologically, hand-tool relations are one of the most significant for our corporeal schematics. As Amparo Lasen insightfully comments, "[t]he way mobile phones are held and touched is one of the aspects that make this relationship different to other ICT devices. The attachment to mobile phones is revealed by the transformation from being an object always at hand to being almost always in the hand and close to the body" (Lasen 2004). As Manovich (2006:2) points out, the making apparent of mobile phone specificity to the consumer means that the interface is no longer transparent but treated in itself as an event. Indeed, using one's mobile phone becomes an aesthetic, corporeal, and meaningful experience; the phone becomes a sensorial whole of textures, colors, lines, materials, movements, and sounds (3). Thus participants became attached to their mobile phones both because they "like[d] the look of it," and were able to personalize the visual aesthetics of the interface with screen savers, wallpaper, skins, and menu structures. Our sample often described having an intimate, tactile familiarity with their phones: "I like to play with my phone when it's in my hand" or "I like to touch the screen—it has a nice feel."

Moreover, the mobile phone becomes an incorporative aspect of the hand while at the same time demanding particular modalities of visual perception. In her work on the biomechanical relationship between the hand and mobile screen device (MSD), Heidi Rae Cooley (2004:137) describes *tactile vision* as a "material and dynamic seeing involving eyes as well as hands and MSD." Our participants claimed dexterity in haptic mobile phone skills such as texting one-handed without looking, while walking or driving—and during interviews several informants proudly demonstrated their adeptness. These activities demand a sensory knowing-ness of the fingers that correlates with what appears on the small screen. Cooley (145) gives the name *fit* to describe the unique relationship between the hand(s) and the mobile screen device, wherein the screenic vision always involves collaboration with the hand or hands: "A material experience of vision results as hands, eyes, screen and surroundings interact and blend in syncopated fashion." In this context, ocular vision is subsumed into a more holistic corporeal vision, by means of which the material contours of the mobile phone, and the haptic relationship between hands, eyes, and device are called into play.

Thus, while it is possible to describe the broad-spectrum nature of screens, our study has shown that the specific phenomenology of screens—in terms of their functionality, size, sensory and somatic requirements, patterns of use, and behavior—must also be accounted for. As theorists such as John Ellis and Chris Chesher (2004) point out, conventions of seeing are not somehow innate, but culturally specific and materially contextual; thus, each "new visual technology emerges with its own conventions—its own 'structures of feeling' and concomitant mechanisms of attraction and counter-distraction." As we have suggested, the particular screen-ness of mobile phones is manifested by how the device enacts visual, haptic, and acoustic incursions—both separately and combined—into our corporeal schema, and so demands variable and oscillating modes of somatic involvement.

Conclusion

In this chapter, we have brought the insights of cultural phenomenology to bear upon an interpretation of the mobile phone as a material and corporealized object. In particular, we have focused on how the device is apprehended as part of the body, how mobile media work to reconfigure our corporeal schematics more generally, and how the mobile screen and contours of the handset elicit a haptic or tactile vision that confounds more conventional regimes of screen viewing. That is, we have argued that the mobile phone has important and ongoing implications for our being-in-the-world. Current trends indicate that the mobile phone is already far exceeding its role as a communication device and becoming a complex new media interface. At a conference several years ago on Urban Screens, Maria Stukoff stated that next generation mobile phone users would be primed to both generate their own made-for-mobile content, and engage "with cinema on-the-go, mini galleries and cultural information via their hand-held devices" (Stukoff 2006). While in Australia such usage may still be relatively uncommon, numerous recent studies on mobile phone use—particularly in countries such as Korea and Japan—have shown the accuracy of such forecasts. The mobile phone is thus clearly emerging as a distinctive multimodal media screen interface in its own right, with its own emergent and distinctive perceptual ratios and techniques. As these developments sediment and proliferate, we must continue to critically interpret mobile media forms in terms of their significance for both our individual and collective corporeal schematics, and the ever-changing dynamics of material culture.

References

Alasuutari, Pertti. 1995. *Researching Culture*. London: Sage.
Chesher, Chris. 2004. "Neither Gaze Nor Glance, but Glaze: Relating to Console Game Screens." *Scan Journal of Media Arts Culture*, 1. Available at: http://scan.net.au/scan/journal/display.php?journal_id=19
Cooley, Heidi Rae. 2004. "It's All about the Fit: The Hand, the Mobile Screenic Device and Tactile Vision." *Journal of Visual Culture*, 3:133-155.
Csordas, Thomas J. 1999. "Embodiment and Cultural Phenomenology." Pp. 143-164 in *Perspectives on Embodiment: The Intersections of Nature and Culture*, edited by G. Weiss and H.F. Haber. New York: Routledge.
Freccero, Carla. 1999. *Popular Culture: An Introduction*. New York: New York University Press.
Gatens, Moira. 1996. *Imaginary Bodies: Ethics, Power and Corporeality*. London: Routledge.
Goggin, Gerard. 2006. *Cell Phone Culture: Mobile Technology in Everyday Life*. Oxon: Routledge.
Haraway, Donna. 1991. *Simians, Cyborgs and Women: The Reinvention of Nature*. New York: Routledge.
Heidegger, Martin. 1977. *The Question Concerning Technology and Other Essays*. New York: Garland.
Hume, Tom. 2005. "Casual Gaming." *The Feature*, 8. Available at: http://www.thefeaturearchives.com/topic/Gaming/Casual_Gaming.html
Ihde, Don. 1993. *Postphenomenology: Essays in the Postmodern Context*. Chicago, IL: Northwestern University Press.
Introna, Lucas and Fernando Ilharco. 2004. "The Ontological Screening of Contemporary Life: A Phenomenological Analysis of Screens." *European Journal of Information Systems*, 13:221-234.
Lasen, Amparo. 2004. "Affective Technologies—Emotions and Mobile Phones." *Receiver*, 11. Available at: http://www.receiver.vodafone.com/11/
Manovich, Lev. 2001. *The Language of New Media*. Cambridge, MA: MIT Press.
——. 2006. "Interaction as an Aesthetic Event." *Receiver*, 17. Available at: http://www.receiver.vodafone.com/
Merleau-Ponty, Maurice. 1962. *Phenomenology of Perception*. London: Routledge and Kegan Paul.
——. 1964. *Signs*. Chicago: Northwestern University Press.
——. 1968. *The Visible and the Invisible*. Chicago: Northwestern University Press.
Okabe, Daisuke and Mizuko Ito. 2005. "Personal, Portable, Pedestrian Images." *Receiver*, 13. Available at: http://www.receiver.vodafone.com/13/
Pink, Sarah. 2007. *Doing Visual Ethnography*. London: Sage.
Robertson, Toni. 2005. "The Stamp of Movement in Human Action." Working paper. Available at: http://research.it.uts.edu.au/idwop/downloads/RobertsonMovement.pdf
Stukoff, Maria. 2006. "The Mobile Urban Screen." *MobileBox*. Available at: http://mobilebox.typepad.com/game_design/2006/08/the_mobile_urba.html
Weiss, Gail. 1999. *Body Images: Embodiment as Intercorporeality*. New York: Routledge.

11
A Grounded Theory Approach to Engaging Technology on the Paintball Field

Ariane Hanemaayer

This chapter introduces the reader to the use of grounded theory in the realm of material culture and technoculture studies and focuses in particular on *generic social processes* of technological interaction. Grounded theory research (Glaser and Strauss 1967; Strauss and Corbin 1998) builds its concepts through data and information gathered from the participants' perspective, own words, and actions. In grounded theory there is a dialectical relationship between theory and data, as the information gathered from participants shapes the inductive framework of the study and guides the refinement of the initial research questions. This method has theoretical implications for a transcontextual understanding of activity in everyday life (see Prus 1996); as research is gathered from various ethnographic inquiries, concepts can be tested across various contexts for relevance and viability.

Grounded in Chicago-style symbolic interactionism and in the tradition of analytic ethnography, this writing takes on an interpretivist and realist approach to the study of technology and everyday life. To explain the significance of the concept of generic social processes I will draw data from an ethnographic study on the activity of paintball.

Analytical and Methodological Frame

The "essence" of technology lies in human interaction. The ways that people engage technics or devices is through intersubjectively accomplished lines of joint action. Taking into consideration the intersubjective, multiperspectival, reflexive, relational, negotiable, and processual aspects of people's everyday lives, symbolic interactionism provides a framework for understanding technological endeavors as human group life in the making (see Vannini chapter five this volume; see also Prus 1996, 1997; for further discussion of these terms and assumptions).

Besides the cited works by Strauss, notable contributions to the methodological perspective outlined here come from American sociologist Herbert Blumer. Blumer (1969) outlined the necessity to do empirical research that both explores and inspects the ways that people do things in everyday life. Through research topics, subjects, and settings, Blumer proposed a descriptive method for understanding the ways that people do things in particular contexts. In his words, "What is needed is to gain empirical validation of the premises, the problems, the data, their lines of connection, the concepts, and the interpretations involved in the act of scientific inquiry. The road to such empirical validation does not lie in the manipulation of the method of inquiry; it lies in the examination of the empirical world" (34).

Such methodological orientation of concept-driven, inductive, analytical ethnography is also the foundation for the development of grounded theory. Glaser and Strauss's (1967) classic work on grounded theory proposes to bridge the gap between speculative theory and ideographic research accounts. While ethnographic accounts that not only inspect and explore but also test the ideographic validity of premises and concepts is important to the development of theory that is empirically viable, there also needs to be a larger picture that develops a more generic understanding of everyday life. Glaser and Strauss propose to develop concepts that are relevant to more than one realm of activity, but that are also empirically viable through the words, actions, and experiences of everyday members. One such tool is that of generic social process. Prus (1996:142) defines it as "[t]he transsituational elements of interaction to the abstracted, transcontextual formulations of social behaviour. Denoting parallel sequences of activity across diverse contexts, generic social processes highlight the emergent, interpretive features of association. They focus our attention on the activities involved in the 'doing' or accomplishing of human group life."

Generic social processes are developed by comparing and contrasting the experiences from one research endeavor with other concepts and experiences developed in other ethnographies and empirical data. In this way, the research questions of a study are answered through words of the participants themselves, but the concepts employed by a researcher are refinable, testable, and viable across various realms of activity. It is in this way that ethnographic research moves beyond particularism and becomes more analytical, that is, transcontextually relevant.

As I discuss in depth later, for this study, interview and field note data were thematically organized by a set of generic social processes set forth by Prus and Mitchell (2006:11) in an attempt to understand the multiple ways in which individuals may possibly interact with regard to technology. These processes include directly Engaging Instances of Technology (ITech), making contact with ITech, achieving familiarity with ITech, encountering enacted limitations of ITech, acquiring procedural technique with ITech, assessing experiences with ITech, comparing options of ITech, modifying/customizing/personalizing ITech, discontinuing usage of and dispensing with ITech, seeking alternatives to/returning to earlier ITech, and devising other uses for preintentioned ITech. These processes—as it should be obvious by their generic-sounding character—are believed to be similar across different contexts and thus research sites.

Background of the Present Study

Over the past 20 years, the paintball game has not only become an increasingly popular recreational activity, but it has also developed into an international competitive event. Competition here unfolds as a war. Players work together in teams as well as by playing solitary opponents (depending on the particular game structure). Players use "markers," or paintball guns, as a device for shooting small, round capsules of paint at one another. The participants in this study played in a variety of paintball activities: (1) Speedball; (2) Recreational Paintball; and (3) Scenario Paintball. Each of the activities is ruled by different guidelines, played with various objectives and point allocations.

Whereas the implication for continued participation differs in any paintball game when a player is "shot" or "hit" with a paintball, there exists similarities among the processes that allow for this research to explore it as a coherent and overarching subculture: (1) the setting has predetermined boundaries for game-play (the "field" where the action is to take place); (2) there are objects on the field that serve as obstacles, as facilitators, or as screens to participants; (3) there are rules that outline acceptable and unacceptable action as well as guidelines that determine how the game is won; and (4) all players use paintball guns[1] and wear a mask to protect their ears and face from the paint. These four similarities in the formation and unfolding of the paintball games themselves allow one to coherently frame paintball as an overarching technoculture, that is, a (sub)culture marked by a distinct and idiosyncratic engagement of technics and techniques.

Design plays an important role in the paintball technoculture. As is to be expected, manufacturers of technological equipment consumed by paintballers take into serious consideration how various pieces of equipment are to be used and played with. For example, guns are produced and targeted specifically for speedball and the need to shoot a lot of paint quickly, though guns for beginners attending recreational paintball also exist. The various "props" (cf. Goffman 1959) used by the participants—for example, to hide (e.g., behind camouflage, bunkers), shoot others (e.g., paintball grenades, guns, or "markers"), structure the game (missions, beginning bells that signify the start of game time), and perhaps more importantly to play their social roles effectively—are central to the performance of the activity. The paintball world has an entire line of products directed at the various needs of the players and game. As with other sub/techno/cultures, the acquiring of this equipment has an emergent aspect in relation to the level of competence at playing the game (e.g., certain products are made for players of higher competence of play than others), as well as implications for a player's identity management on the field. Notably, in my fieldwork experience, much of the discussion during paintball fighting was focused on what gun someone was using that particular day, what strategies they used to capture the flag, or, for example, how a player might fix a technical malfunction with their gun. In sum, not only are players directly engaging technological tools in the here-and-now aspects of engaging in the activity, but their participation in the subculture more generally is technologically focused (for a similar argument see Roustan this volume).

The present ethnographic study was conducted using in-depth interviewing with four male participants, and especially hundreds of fieldwork hours and subsequent discussions I had with players at a local paintball field. Fieldwork consisted of participant observation, and notes were taken during and after game-play and field discussions. Interviews varied in length from two hours to six hours, and were open-ended, but guided by the particular interest in the generic social processes outlined by Prus and Mitchell (2006) on using technology. Interviews were transcribed and concepts were organized using the subprocesses employed by Prus and Mitchell.

Once I wrote up my field notes and transcribed and organized my data into themes I began to look for the generic elements of engaging technology on the paintball field. Using the framework of generic social processes outlined in Prus and Mitchell I assessed the viability of the processes for the data I collected. Quotations and field note observations were selected to best illustrate a particular subprocess of engaging technology. While ethnographic work in the tradition of grounded theory is generally very detailed and focused on developing concepts, in this short chapter my main goal is illustrative of the genre of grounded theory, the concept of generic social processes, and thus more pedagogical than analytical in the strong sense of the word.

Using Technology

Generic social processes of the kind examined here are organized through increasing involvement with technology, such as encountering technologies, developing a sustained understanding of technology, disengaging from and possibly reengaging particular instances of technology through the lens of a career contingency model of group life processes. These processes are by no means the only ways in which people engage technology directly. Instead, they merely offer a framework for organizing ethnographic materials and for conceptualizing technology in everyday life.

To situate the use of technology within the realm of the paintball subculture, let us define the technological aspects of participation in this activity: material objects (or technics) and strategies of use (or techniques). Technics include such things as equipment that is worn by the players (such as a mask or "pod pack")[2] or used during play (such as paintball guns, also known as "markers," or paint-grenades).[3] In addition, other materials may be found on the field, such as the obstacles, "bunkers," or buildings built to simulate a city center. Strategies or techniques may be defined as a series of actions that people mindfully and attentively develop to use technics as well as how people prepare for, implement, adjust to, and resist the activities of others (including themselves, their team members, or opponents) to achieve their own desired objectives. These techniques of play are emergent in nature, as people may continuously reengage and/or readjust to their anticipated objectives and to the reactions of others.

Instances of technology have emergent temporal qualities, as a participant's career of engaging technology is assessed and reassessed with each action. Mindful of the point that instances of technology may have "(a) a multiplicity of meanings that people may associate with particular [objects] and (b) [a] somewhat related multiplicity of operational standpoints with which people may engage specific objects" (Prus and Mitchell 2006:9), instances of technology are based on interaction and varying definitions of the situation at hand.

Objects, on the other hand, may be material (e.g., equipment, clothing), social (e.g., team members, friends), or abstract (e.g., techniques, principles) (Blumer 1969). Objects have no inherent meaning, as their meanings are achieved and modified through interpretation and an emergent process of social activity (Mead 1938). Material objects indeed have an ambiguous quality in that they acquire meaning in the process of their use, while, to some degree, remaining constant in their material properties. Thus, instances of technological interaction with the same object may differ over time according to their envisioned use, especially if an adjustment has been made in light of the "resistances" objects offer (cf. Mead 1938). With regard to this, Prus and Mitchell (2006:9) argue that "like other aspects of the situation that the participants may define in one or other ways, the people involved may later decide that particular objects failed to offer the advantages they had anticipated." Meanings assigned to objects and techniques change over time.

In the interest of space for this brief chapter, I choose only a few subprocesses from the list of generic social processes generated by Prus and Mitchell (2006): (1) making contact with ITech; (2) achieving familiarity with ITech; (3) encountering enacted limitations of ITech; (4) modifying.

customizing, personalizing ITech; (5) discontinuing usage of, dispensing with ITech; (6) seeking alternatives to, returning to earlier ITech.

Making Contact with ITech

By framing how people use devices through a concept of career contingency, analytic ethnographers can further develop how people become initially involved (Prus 1996, 1997) with particular objects. People may seek out objects to accomplish specific tasks, just as people may be recruited by others into using particular objects. Seekership and recruitment, however, are related to people's desired objectives. People develop desired intentions by making initial engagements with instances of technology.

With that notion of a desired accomplishment in mind, people may not only seek out enabling devices, but they may also be recruited into using particular items to achieve their accomplishments. Or, they may inadvertently find a way to resolve their search for a desired technique.

In my fieldwork, players often recounted their searches for paintball guns. Players often had notions of what they wanted their guns to look like, act like during play, and so on. However, making contact with a gun that fulfilled their anticipated desires was sometimes problematic:

> I grew up doing a lot of hunting and fishing. And I knew how a real firearm should work. And much to say that, it really governed how I wanted a paintball gun to shoot. Ok, because I knew a real firearm shoots very straight and accurate and whatnot.

This participant had been involved in paintball for over 20 years, so when he started to play paintball, guns were not as accurate as he had imagined or hoped for.

Other players also described using Internet forums as a source for finding out information on available equipment and feedback on its performance from other users. However, this kind of searching can be problematic, as the authenticity of the data had become questionable for some players.

> The problem is that now that paintball has become more commercial, more mainstream, all companies put people on there to write biased reviews, or people that haven't even used it are writing reviews. So it's kind of useless now or at least in my mind.

These online forums can sway or influence people to buy products, if players' desired criteria are met. Sometimes peers or store merchants can recruit players into making purchasing choices. One player described how a retail agent, who was a tournament player, convinced him of the desirable uses of another paintball gun. So while players may know what they want, they may also get influenced and recruited by those around them to make certain purchases.

The ways that people make contact with instances of technology is through a process of first deciding what they would like to accomplish, or the desired qualities or characteristics of a particular device, followed by finding something that they can use to achieve that end. People may seek out technical devices on their own, be recruited into using various technics through influence work, experience a sense of closure, or inadvertently come into contact with them. More data that develop how people learn about or come into contact with various techniques of play or acquire strategies could provide depth into how people make contact with instances of technology more generally. Also, another venture for analysis could develop the ways that people nurture desired outcomes and go about finding sources for formulating plans and accomplishing them.

Achieving Familiarity with ITech

After initial contact people begin to use particular devices and they progressively achieve a sense of competency, adeptness, and familiarity. People become familiar with objects in an emergent fashion, through repeated instances of use. As Vannini explains in chapter one, this process is also known as domestication. As people use things, they consciously reflect on the similarities and differences of the objects, and make adjustments according to their desired use of the objects. For example, Puddephatt (2003:268) notes that chess players often have "certain memorized rules of thumb, operating systems, strategic approaches, particular problem solutions" to define situations while playing. While people may be knowledgeable about certain objects, once they enable these devices and put them to use they may not be the masters that they anticipated they would be. Instead, engaging devices can sometimes be problematic as people deal with setbacks and develop customs for directly engaging them.

Many paintball players described the process of achieving familiarity with both materials and strategies as a series of "trial and error." Players described practice as an effective way to learn to use a particular object, although this may be, at times, problematic. As people engage various materials, they may become mindful of certain characteristics. Most players described competency in relation to taking into account the environment, such as wind speed, and manufacturing specifics, such as ball speed, barrel design, and so on. While using the objects and taking into account these and other characteristics, people develop strategies for coping with limitations, and these adjustments lead to higher confidence in using the objects. For example, consider the following quotes from one of my interviews:

> Ultimate Air snake bunkers are higher than they are like a foot and a half, two feet, you can crawl easier and you don't have to be as low to the ground. Whereas Sup Air is a very short, about only a foot maybe. So you have to really stay nice and low so.

> If, whenever possible, you can get, with the wind at your back, you'll get another 50 feet at least because now you're shooting with the wind and it's actually carrying the ball...I've taken like twenty shots to see which way they're [paintballs] going to go and I adjust accordingly, and I drop them in on the [opponent].

As people intensify their involvements with particular objects, they seek to develop a sense of familiarity and competency for accomplishing instances of technology. However, when they use objects over time, they will encounter certain limitations. Let us then explore how people deal with this.

Encountering Enacted Limitations of ITech

Like any human action, instances of using technology come with setbacks and limitations. As they become competent at using particular devices or strategies, people may also learn to deal with unexpected changes, failures, and adjustments frequently when using technics of various kinds. Design limitations are encountered whenever people act toward objects and do not accomplish their desired outcomes. Limitations may emerge at any time. In turn, these limitations may lead to the development of disaffections with particular objects or techniques or to the development of new adjustments or techniques.

For example, following initial contact with an object, paintball players may define it as limiting. In buying a paintball gun, for instance, people may take into account the ease of seeking repairs should they need them in the future, and discover that their desired tool may not have accessible replacement parts. Therefore, when players discover that a particular gun is not accessible, they may abandon that pursuit altogether and choose to use another gun.

In addition to limitations of availability for repairing or acquiring objects, paintball players may also define the features about a particular gun they do not like after using it. This knowledge about the limitations of the equipment may lead players to avoid using it again. For example,

I don't like the accuracy, the action, the gun is like kaboom kachunk and it goes all over the place. I don't like that at all.

Other encounters with limitations may happen while directly engaging play with a particular piece of equipment on the field. Sometimes, as players adjust to particular limitations and failures of equipment in game-play, they attempt to fix the malfunction, but they are not always successful.

Players are concerned not only with limitations affecting the desired use for their materials, but also with the implications that malfunction may have for the game strategy and unfolding events of play. Enacted limitations are defined in the instance of interaction, but may have emergent qualities, as a series of limitations, each with implications for the accomplishment. In addition to equipment failures, occasionally, when people use props, they discover the limitations props have for the players' desired use. Consider, for instance, the following:

CXBL [an international game competition] last year [used] ultimate air [brand that makes inflatable bunkers]. The small dorito was up to my knee so you can't hide behind, so it's a useless prop on the field.

These limitations may occur at any moment during their direct use. When people experience limitations with their material objects they may choose to disengage them entirely, no longer seeking to use them, or they may assess the severity of the problem and make adjustments and attempts to solve the failure. Limitations may also be defined as the result of resistance from others. As people try to execute techniques for achieving desired ends, the actions of others may thwart their attempts to be successful. Limitations within the instances of engaging technology, then, are emergent events that are contingent on the material properties of the object, be it material (equipment), social (other players), or abstract (game rules). The relation between objects and people is dynamic, and the activities of objects—in this instance, limitations—may alter the lines of action of the people using those objects (Mead 1938).

Modifying/Customizing/Personalizing ITech

Although techniques are undoubtedly shared across people, and within a culture, people may also develop their own individual style. These modifications are developed through people's assessments of their experiences with technology, and how they acquire familiarity for using objects in a particular way.

Also, as people take their experiences with, and desires for, using objects into account, they may develop and put into action particular strategies of their own. Important to note here is that although aspects of group interaction such as rules in a game are intersubjectively negotiated, the human ability for spontaneous and creative thought is also possible. Although options may be encountered or shared by the group, human agency is central to understanding this process. The ways that people modify, customize, and personalize their uses of objects may be done along with the group or be developed by an individual on his/her own.

In paintball, modifying, customizing, or personalizing technology could be related to upgrades meant for improving the equipment's performance, or may have to do with the ways that people changed strategies during play, and, subsequently their own lines of action, to engage objects for their desired accomplishments. While changes and modifications in strategy are frequent in paintball, here I can only focus on the techniques that people use to customize their equipment for a better performance. Consider, for example, the following:

> I've probably tried everything out there that you could possibly try. The first guns I had was basically pretty crude aluminum barrel that I polished them. I would get brasso and polish them right up. I found the paintballs would fly better and straighter because the balls would slide in there.

As this participant makes clear, people make changes and personal modifications based on their own desires. Equipment maintenance is an example of how people may want to change the performance of their materials. Each participant, for example, wanted to change the way they used their guns, so they sought after their own methods of improving this quality.

People make modifications, customizations, and personalizations in relation to the ways that they desire to use a particular object. These changes and adjustments are made when taking into account techniques for engaging the object. While some paintball players make modifications that were aesthetic and not always in relation to performance, people nevertheless make changes to objects to satisfy a personal desire. These changes may emerge through group interactions, but each person's agency allows them to assess and develop their own desires through engaging objects in the instances of their use.

Conclusion

Approaching the study of technology in human group life as an emergent, socially, and meaningfully enabled generic process of using devices through grounded theory and the theoretical framing of Prus and Mitchell (2006) provides analytic ethnographers with a list of viable concepts and processes to compare and contrast other realms of activity and to build a more generic understanding of everyday life. Generic social processes seek to create a transcontextual, transsituational, and transhistorical conceptual-ization of human activity. By defining human group life in process terms, generic understanding can emerge, be empirically tested, and emergently extend human knowing and acting.

This chapter demonstrated how analytic ethnography can provide viable ways to examine and inspect human group life using grounded theory. Grounded theory allowed this research to approach the study of how people use technology from a "bottom up" perspective, examining and detailing what people *actually do* when playing paintball. Offering researchers a methodology that lets conceptual themes emerge through fieldwork and repeated analysis, grounded theory uses the words and actions of the participants to construct a viable understanding of their lifeworld.

When doing ethnographic work, researchers are encouraged to develop themes from the participants' own words and actions, build upon concepts from research in other realms of community life. While some ethnographic projects may seem drastically different in their forms of activity, there may be generic elements that can be empirically tested in other contexts of everyday life, such as the themes mentioned in this chapter. To further assess the current themes on how people use technology, research can be pursued in other realms of human activity, testing whether these concepts are indeed viable in other human ventures. Through ethnographically grounded analysis, researchers can continue to compare, contrast, test, and adjust transcontextual concepts, or generic social processes, in human group life.

This methodology and level of in-depth, grounded analysis will lead research about human group life to be more generic in nature, and will provide human scientists a broad as well as deep understanding of what and how people do things.

Notes

1. Equipment that shoots pellets of paint using either compressed air or carbon dioxide.
2. Pods are plastic containers used to hold extra balls of paint to provide hasty refills during game-play. Pod packs are vests that have holsters to insert numerous pods for refilling.
3. A plastic casing that looks like a military grenade that sprays paint when hit against something hard.

References

Blumer, Herbert. 1969. *Symbolic Interactionism: Perspective and Method*. Berkeley: University of California Press.
Glaser, Barney and Anselm Strauss. 1967. *The Discovery of Grounded Theory: Strategies for Qualitative Research*. Chicago: Aldine.
Goffman, Erving. 1959. *Presentation of Self in Everyday Life*. New York: Doubleday.
Mead, George. 1938. *The Philosophy of the Act,* edited by Charles W. Morris. Chicago: University of Chicago Press.
Prus, Robert. 1996. *Symbolic Interactionism and Ethnographic Research: Intersubjectivity and the Study of Human Lived Experience*. Albany: State University of New York Press.
——. 1997. *Subcultural Mosaics and Intersubjective Realities: An Ethnographic Research Agenda for Pragmatizing the Social Sciences*. Albany: State University of New York Press.
Prus, Robert and Richard Mitchell. 2006. "Engaging Technology: A Missing Link in the Sociological Study of Human Knowing and Acting." Paper presented at the STS Colloquium Series, Department of Science and Technology Studies, Cornell University, Ithaca, October 30.
Puddephatt, Anthony. 2003. "Chess Playing as Strategic Activity." *Symbolic Interaction*, 26:263-284.
Strauss, Anselm and Juliet Corbin. 1998. *Basics of Qualitative Research: Grounded Theory Procedures and Techniques* (2nd ed.). Thousand Oaks, CA: Sage.

PART 3
ETHNOGRAPHIC STUDIES

12
What Gardens Mean

Chris Tilley

I want to start off by telling you a tall story. Geoffrey, a novice anthropologist and inhabitant of the Trobriand Islands, had recently been inspired by a reading of Malinowski's monograph *Coral Gardens and Their Magic* (1935). Searching for a PhD topic, he suddenly had a brilliant idea, quite audacious in its simplicity and power. He would go to Malinowski's adopted home country and undertake a study of English gardens and their magic. Turning the tables of colonialist discourse and the exotic and peculiar practices of English savages might constitute a very interesting study indeed. From what he could understand from a study of the few books available to him, the English seemed to take their gardens very seriously. Rudyard Kipling, one of their famous poets, had even gone so far as to state that "the English are a nation of gardeners." From Web sites he learned that some of the most beautiful and sophisticated and grandest gardens in the world were to be found in England. However, the anthropological tradition and subaltern studies do not record these "refined gardeners" but ordinary people and what they grew in their backyards. To his surprise Geoffrey found that scarcely anything other than anecdotal had been written about ordinary English gardeners and their gardens despite a voluminous literature on garden history discussing the gardens of the rich and the powerful.

Geoffrey took a small plane to Port Moresby. From there he flew to Sydney and on to London. The plane circled over the great city in arcs gradually getting lower and lower. His first glimpse of the English garden was through the plane window. London was huge but very, very green. There were massive green spaces—presumably community dancing grounds—at the center of the city and then millions of little brick boxes standing next to each other in rows, a paved street between them, with small patches of green at the back and the front.

How well fenced these bits of green space were! The English clearly had some kind of obsession with borders and boundaries and making themselves distinctive. From a reading of the work of the English anthropologist Marilyn Strathern (1988), Geoffrey had come to believe that he himself should be characterized as having a special kind of agency: that associated with being a "dividual." Here in England he could expect to meet an entirely different kind of person with the rugged kind of agency associated with being an individual. While he and his family did not really own the products of their labor—all was, after all, process and relationality in the ever-shifting and kaleidoscopic contexts of production and exchange—he could expect to find an entirely different English attitude, the essence of which is alienation.

It was a warm Saturday evening in the early English summer. As the plane descended closer and closer to the city Geoffrey noticed a pall of smoke. This seemed to be emanating from the countless small green patches at the backs of the houses. All the English seemed to be engaged in burning something in their gardens on this balmy summer evening. Exiting from Heathrow, amidst the diesel fumes, he sensed something else very familiar to him: the smell of burning pork.

Geoffrey ardently wanted to give voice to his English informants. His key research question was why so many of them seemed to care so passionately about these little patches of land and worked so hard in them. How could they regard this very peculiar economically unproductive and indeed wasteful behavior as in any way normal? Both men and women worked in the garden; people from all walks of life and classes also worked in the garden: mechanics, hairdressers, and barristers—all seemed to share this interest. Surveys had shown that gardening was by far the most popular outdoor leisure activity in London (Bhatti and Church 2001). The shelves of the bookshops were groaning with instruction manuals and books about flowers and vegetables. TV gardeners were celebrities and held in great esteem: these were the masters of English garden magic.

It proved quite difficult for Geoffrey to enter the English garden at first because of all the physical and social boundaries at work. The inhabitants were quite suspicious of his motivations. They did not fully appreciate the benefits of being studied. There was absolutely nowhere that he could set up his tent at the center of any English street, town, or village.

The English professed to have a great love of plants but this manifested itself in a rather peculiar manner. At the heart of the English classificatory logic was a distinction between garden plants and weeds. Most of their own ubiquitous indigenous species such as nettles, brambles, daisies, buttercups, dandelions, wild roses, and so on are regarded as weeds. They are violently destroyed by various mechanical and chemical devices. They like to fill their gardens instead with wild plants originating from every corner of the globe: South Africa, China, Japan, North and South America to mention just a few more distant countries, or closer to home, the European Alps. They tend and cultivate plants of other countries while despising their own. Most do not grow anything at all that they can actually eat. This is because they regard their gardens as outdoor living rooms and occasional kitchens with lawns as carpets, hedges as walls, flower borders filled with trinkets and other decorative devices, and the barbecue as cooking fire. Some of the plants they cultivate are regarded as sophisticated, others as garish or vulgar according to class and esoteric garden knowledge. In order for their plants to grow well they have to look right.

Some gardens are replete with small effigies termed gnomes that may be seen engaged in a wide variety of postures and activities (sleeping, smoking, fishing, meditating). There are often small statues of wild animals, for example, foxes, deer, badgers, and rabbits, which are otherwise ruthlessly chased and killed in the countryside. The grass lawn is a particular fetish for some. It will often be shaved almost bare of vegetation two or three times a week and the English engage in an extraordinary magical practice of mowing by means of which stripes are made to appear only to vanish again after a few days.

A large number of the English claim that they love their gardens and find being in them very therapeutic and relaxing. Yet most of the time they are constantly working in them, sometimes exhausted. Occasionally they may sit in their gardens but the repulsive sight of the enemy—an English weed—will spur them into action once more. They constantly watch and are concerned with what their neighbors are doing and growing. Some of these neighbors growing particularly tall trees along their garden boundaries are popularly termed "neighbors from hell" and appear in TV programs.

After a few months of study Geoffrey realized that he had indeed struck the goldmine of anthropological research. The English, their social and political relations, their entire cosmology of life was indeed rooted in their gardens in just the same manner as Malinowski had claimed that Trobriand identity was rooted in theirs.

* * *

I have started off this chapter, you may have noticed, with a classic phenomenological strategy, the bracketing of experience. This is a refusal to accept that there is anything normal about gardens and gardening that, it might be added, is somewhat difficult for me personally as I have been gardening since about the age of ten. This chapter attempts to outline what gardens mean in contemporary England. From interviews with sixty-five English gardeners, which were undertaken in their gardens during the summer months of 2002-2004,[1] I outline here ten of the main reasons why gardening has such significance for many people. Of course there are many other reasons, but from the interviews and watching people garden, I deem these to be the most important. The text deliberately moves back and forth between the deeply personal to a much broader cultural understanding of gardening practices. Gardens, I maintain, are small places that are a source for contemplation on big issues.

Gardens as Material Metaphors for People's Lives

John had spent much of his life as a dustman driving the rubbish lorry. He rents a small local authority-owned terraced house. There is a small patch of concrete, only a few meters wide, at the front with some empty concrete tubs. At the back of the house there is a narrow rectangular space going up a steep slope with a central concrete path with steps and an area of lawn either side.

Figure 1: John's Front Garden

When I visited John, his[2] garden was virtually dead. Apart from the grass there was nothing growing in it. There appeared to be little to talk about (Figure 1) but John gave me some photos of how the garden had been: lots of tubs at the front and the back full of geraniums and busy lizzies (Figure 2). His wife had been very fond of bright flowers, especially the pink ones, but the previous summer she had died. In the autumn John put the plants in the greenhouse to protect them from the frost. He forgot about them and they all died. He had never planted anything since.

But John was still maintaining an allotment down by the churchyard. He was growing only two things: runner beans and potatoes, far more than he could eat. He described himself as a proper gardener of the "old sort." He gave most of his produce away to a network of neighbors, family, and friends. The potatoes and beans were a material medium through which he maintained relationships with others. He liked to sit in his allotment, as he had always done. This was his place. His house and garden, his dead wife, were out of sight and out of mind.

Figure 2: The former appearance of John's front garden

Susan lives in a big rural house, part of a small farm. She had spent most of her adult life living in Hong Kong. There she grew brightly colored plants: deep reds and strong yellows. Unhappy in her marriage she had got divorced and moved to England. Here in her new English home with her new husband, by contrast, she thinks soft pastel colors—such as pale pinks and blues—are more pleasant. She told me that she was always moving the plants in her garden around to find the very best place for them. She kept telling me, "if a plant is unhappy, you've got to move it"

Ann and Lesley lived in a modern four-bedroom house on a suburban housing estate with quite a small garden. They have two young children Ann looks after at home. To supplement the family income Ann works as a professional child caretaker. Most of the garden is full of play equipment. When I talk to them about the garden Ann and Lesley disagree and argue about almost everything. Ann likes flowers, but Lesley is interested only in vegetables he grows in containers because of the lack of space. I ask them to take two photos of their garden for me. Ann chooses a small patio area with flowers at the front, whereas Lesley chooses to photograph one of his vegetable containers at the back. Ann's second photo is of the only part of the back garden without play equipment (Figure 3).

Figure 3: Ann's photo of part of her back garden without play equipment

Lesley takes a photo of all the play equipment filling up and dominating the garden (Figure 4). This was almost certainly a photo taken with irony. Shortly after I interviewed them they split up and are now divorced.

When I visited Neville in his garden he had just planted a sapling. He was in his eighties. He said to me, "well, I'll never live long enough to see it grow up into a tree, but you can't think like that you know. You've got to carry on!"

Every garden, like artifacts (Hoskins 1998, 2006; Woodward this volume), has its significant stories, memories, and its biographical associations. Gardens change as people and their circumstances change. Each is intertwined and one cannot make sense of the garden without knowing the person and vice versa.

Both gardens and persons have their lifecycles, and gardens are often powerful material metaphors for this (Tilley 1999). What is appropriate or possible at one time is not at another. So why do people garden?

Figure 4: Lesley's photo of the back garden

A Love of Plants or Care of the Self?

Over and over again you will hear in the popular media the claim that the English garden because they love plants, have an emotional feeling for plants, and some even have great knowledge of plants. My understanding is rather different. The contemporary English love of plants is really a means of loving themselves and caring for others. The cultivation of the garden is a way in which people care for and cultivate themselves, a technology of the self in contemporary modernity (Foucault 1987). In a culture of mass consumption in which we buy products that we have not produced and whose conditions of production and exchange largely remain a mystery to us, gardens by contrast, are usually things that people make themselves. So gardening is a craft through which many people cope or come to terms with their existential alienation in a world in which the normal condition is estrangement (Ollman 1971). We ordinarily do not know how the things are made, who makes them, or often even where they come from. Gardening is always more than being a creative consumer choosing and personalizing the things that have been purchased, such as the act of arranging furniture in a room (Miller 2002). It is also about creating something in a much more fundamental way. Gardeners may buy plants as consumers but many also cultivate, or create, the plants themselves from seeds and cuttings, and by means of dividing up plants. They may prune, trim, and maintain these plants, create arrangements of flower borders, make paths, arches, trellises, and other structures with their own hands. While we normally do not make the houses we dwell in and often may not maintain or decorate them ourselves, or make the fittings and furniture, people usually make or maintain, add to and change their gardens. By doing so they make themselves.

We have all been accustomed to the term "retail therapy." In a consumer society we make ourselves happy through buying things. The widespread notion of an alienated and fragmented modern consumer self, depending on accumulating more and more things and creating new identities from them (Featherstone 1995; Mackay 1997; McCracken 1988), can be contrasted with the types of identities that may be constructed through gardening. Gardening is fundamentally about the therapy that comes from creating something yourself and the investment of personal time, energy, and artistry (Francis and Hester 1995). Garden work involves investment of work and effort, and so gardening can be understood to be a labor of love. The love of plants is a means of creating both a personal and a social identity through a vegetative medium. The gardener may be a creative consumer of plants from garden centers and so on but he or she is normally also a creative producer and has an intimate knowledge of the thing, the garden that has been made. Growing a tomato from a seed, tending to the plant, harvesting and eating the fruit is an altogether different creative act from purchasing a commodity from a shop and then deciding where to put it and what to do with it.

Thinking and Dwelling

Gardens compared with most of the other things we refer to as "material culture" are relatively both large and durable. A gardener dwells in the garden. He or she is inside the thing itself, inside the garden that he or she has created. As Cooper (2006:81) puts it, the gardener is "copresent" with the garden and is its "cocreator." Thus in a metaphoric sense the gardener is inside himself or herself, in a garden body, underneath a garden skin. The garden cannot be moved without being destroyed. One of the most significant aspects of the garden to most gardeners is that these are outdoor spaces and in this respect is often opposed to the house. You escape from the house and go into the garden. Some gardeners prefer being in their gardens to being in their houses or they say that they can only be themselves, or find themselves, in the garden. Garden work, to many, is pleasurable whereas housework is universally considered to be a chore: doing weeding is always preferable to doing the dishes. Many people like to be outside in their gardens rather than inside in their houses for a number of reasons: the garden feels less confined. It can free up your thought in unexpected ways. It provides a far richer sensorial environment than the house, changing constantly with the weather, the time of day, and the seasons. By contrast, the house and its interior are static and limiting. The garden is a continual source of wonder and surprise and is hence rewarding. One can, for example, observe a rosebud gradually unfolding. Chris took this picture of his garden: fennel and hollyhocks intertwined and growing together (Figure 5).

Figure 5: Chris's photo of part of his garden

What was of particular delight to Chris was that he had planted neither of them. Both were self-seeded and had decided of their own to grow together. The garden in this sense has its own agency that is quite independent of the agency of the gardener (Gell 1998; Hitchins 2006). Sometimes the agency of gardener and garden is regarded as working harmoniously together. At other times the gardener has to fight against the garden to keep it under control. Vigilance is needed or the garden will get out of hand. By contrast things in the house remain pretty much the same. Pictures cannot move themselves, get bigger or smaller, sofas don't sprout from sofas.

The sensory dimensions of the garden are absolutely fundamental to many: feeling the sun on your face, walking barefoot on wet grass, smelling apple blossom, feeling the soil running through your hands, hearing birdsong, seeing a beautiful flower, tasting the fruits and vegetables that you yourself have grown. A favorite garden task for many women was weeding. Several gardeners explained this as being all about getting down among the plants and being able to observe and interact with them more closely. Although almost all gardeners had gloves, they rarely wore them, and as a result, their hands were always black. They liked to touch and feel the plants and feel the soil running through their fingers. The garden should look aesthetically pleasing especially when seen from the house. The pleasant scent of certain flowers was particularly desirable to many; the sounds of birdsong, wind blowing through the leaves, and murmuring of a brook were also quoted by many. All this makes the garden a place of dreams and being in it becomes part of a daydream, a true Eden on earth (Tilley 2006b). Such dreams may, of course, be rudely disrupted by the sound of the neighbor's electric strimmer or jets passing overhead, the noise and fumes of traffic, or a farmer spraying nearby fields with pesticide.

Being in and Caring for Nature

The garden, although made and therefore a cultural artifact, has a life of its own and is therefore thought of by almost all gardeners as being a natural place. People may cultivate plants, but they are not made in the same sense as one builds a house or makes a table. Nature is often equated with the wild: something untouched by human hands. Garden plants that one may buy from a garden center are cultural and consumer products just like any other, but having a life of their own, they are beyond human control. So they may be considered, like domestic animals or indeed human beings, as being "quasi artifacts," somewhere between nature and culture (Latour 2004). However, gardeners do not think in this way. While they acknowledge that they create and maintain the garden, they regard this as being a natural place that gives it its positive value and meaning, in a quite different manner from the house. Compared with the confines and constraints of the domestic interior, to many, being in the garden provides a sense of freedom, indeed escapism. The house is an artifact that may constrain and limit while the garden is full of potential and promise, a place of dreams.

I showed a photograph with trees growing on a carpet of moss to my English informants and asked them whether they liked this garden. Most thought the image was rather serene and beautiful, but it troubled some who questioned whether this could really be a garden. Others suggested it might be part of a very large garden. The issue was the lack of human intervention. In other words, for a garden to be a garden there need to be explicit signs of human cultivation. One of my informants was a strong advocate of wildlife gardens, the idea that has a growing popularity, that gardens can be mini nature reserves (e.g., Kingsbury 2005; Lewis 2003). She was cultivating a wildflower meadow in the English countryside, and her garden was occasionally open to the public under the National Gardens charity scheme. So if this was a wildflower meadow that one might find in those few parts of the English countryside that are not intensively farmed and sprayed with chemicals and fertilizers, was it really a garden? It was a garden only in the sense that she had made it and the meadow itself was only part of a much larger garden that did not have any wild flowers in it apart from English weeds that were removed periodically. It is undoubtedly the case that for most gardeners gardening in a general sense is an expression of caring for nature and the environment hence the growing popularity of organic gardening and regarding the garden as a form of wildlife sanctuary. Yet there are limits to this: most English gardens contain few indigenous English plants, and badgers and rabbits are delightful so long as they don't do what rabbits and badgers do: eat up your lettuces and dig up your lawns!

The Meaning of Garden Labor

When we consider cases such as that discussed above, it leads us to the conclusion that the defining characteristic of the garden is not so much the plants that it contains as the human labor that goes into making and maintaining it. Gardens are by far the most worked over patches of land on the planet. The amount of personal labor invested in them is often enormous. Some gardeners who work spend almost all their spare time in their garden and may not even feel that they can go on a holiday for any extended period lest the garden should get out of control. For some retired people, the garden becomes their life. They live to garden and this is interesting with respect to the agency of the garden and that of the gardener. The gardener may believe that he or she is controlling the garden, but in reality the garden controls him or her who no longer has any freedom or sacrifices this freedom for the sake of the garden. To many gardeners to work in the garden was a fundamental part of finding meaning. Gardening was a way of spending or filling up time, and they would be bereft of gardening. "What else would I do?" they would say to me. According to many, working in the garden made life meaningful and tolerable. The practice of gardening was an existential necessity and the more work that needed to be done, or could be created, the better. When I asked if their garden was finished, I often got a surprised response: "a garden is never finished!" they would say. The independent agency of the garden means that it can never be finished. The plants grow and need cutting back or tending, weeds invade the garden space, fences and hedges need keeping in order. There are always a myriad of tasks that need to be undertaken.

Most gardening is a rather solitary undertaking. Gardeners usually garden on their own. Even if couples are working in the garden at the same time they will usually work independently doing different tasks in different parts of the garden. Even if engaged in the same activity such as weeding, rather than weeding the same border together, they will typically weed different borders. Many gardeners find gardening relaxing and therapeutic precisely because they can escape from others and be on their own. It is an escape from talk and having to be sociable and exchange pleasantries.

Gardening is practical work, work with the hands, and not having to think and intellectualize is why many find it pleasing. In the garden one just carries on, and this needs no further justification. It is an activity that has value in and by itself, an end in itself. No explanations are required. Gardening does not need to be put into words or justified by words. Only TV gardeners, or garden writers, need to do this. It is a stream of unconscious activity that becomes the stuff of dreams. So gardening is an escape not only from sociability, but also from the demands of conscious thought and language. This is why even though the task may involve very hard work it is still regarded as personal therapy. Typical of so many stories I heard from my informants (Kaplan and Kaplan 1995; Lewis 1995), Marjorie, a journalist by profession, told me about having had a nervous breakdown and the manner in which gardening had helped to restore her mental health and zest for life.

A lot of the pleasure of gardening and the meaning it has for people is all about being able to see the results of your own labor: that neatly trimmed hedge, that beautiful camellia that you planted and tended, that well-weeded border. The garden acts as a kind of mirror reflecting the work of the gardener. He or she can see their work materially objectified in it (Tilley

2006a), and they can feel proud of themselves (or unhappy) in the process. The garden is part of them and they are in it. In this sense gardening is equivalent to creation: the ability to give new life, just like being an author or an artist. Some people told me that although they cannot paint with a brush, they attempt to paint with flowers. So the garden is a material objectification of labor, and to be a good gardener is associated with being a good person. In just the same way, a poorly maintained garden reflects badly on those who cannot be bothered to make the effort. In most occupations it is not normally possible to see the direct results of your work—for example, working in a production line in a factory or in an office—because the overall labor process swallows up and envelops individual efforts.

The majority of gardeners would not employ a garden designer even if they could afford to do so. They want to do the work themselves, however poor the outcome might be deemed by others. This is because they want the garden to be an expression of themselves. Generally speaking, the people who do employ garden designers are those who have little interest in, or knowledge of, gardens and gardening, but may want to impress others or have an easy ready-made solution to having to maintain a garden space. Keen and knowledgeable gardeners who employ people to work in their garden only allow them to do tedious and repetitive maintenance tasks such as hedge or lawn cutting. Any creative tasks will be reserved for themselves, although they may often be open to advice and opinions.

Love and the Divisions of Labor

In the garden there is normally a very strong division of labor when couples are involved. Men typically cut hedges and lawns, do heavy digging and "hard landscaping" such as making patios, trellises, arches. Women, by contrast, concern themselves with tending for ornamental plants, planning the garden, weeding, dead-heading flowers, and many other small tasks. Much has been made in the popular gardening media, principally books and magazines, that men enjoy such activities as lawn cutting and hedge cutting because it contributes to a macho male identity, and has to do with a love of wielding, using and maintaining machinery such as garden mowers and hedge cutters than any real appreciation of gardens and gardening (e.g., Anonymous 2001). The results of my research challenge this conventional wisdom. Although a minority of males do indeed love machines and like using them in lawn mowing and hedge cutting "rituals," the majority perform these tasks because their spouses do not like doing them and prefer them to do so. Theirs is a labor of love rather than the projection of a particular form of male identity.

The stereotypical division of labor with men cutting lawns and hedges and doing strimming and heavy digging is maintained and established by their spouses, who are usually the ones to assert that their male partners like performing such tasks. It is interesting to know that out of the twenty-six English males interviewed only two actually stated that cutting lawns or hedges was a favorite activity.

It was often said to me that it was natural that men should do the heavier jobs because physically they were stronger, but this is largely a rationalization because in single person households women often did all these things. There were relatively few cases in which both male and female partners both liked gardening equally as much, spent an equal amount of time gardening, and were together in decision making in the garden.

What went on in the garden was simply an extension of decision making about decorative schemes and so on in the domestic interior (cf. Chevalier 1998; Grigson 2007). Mostly it was women who had the vision and decided how the garden should be, just as they controlled decision making in the home. The male of the household was cast, and frequently cast himself, in the role of a helper, sometimes a somewhat reluctant one. So both the division of labor and tasks appropriate to men and women were usually maintained and defined by women rather than men and were in accordance with their interests. It was they who sent their partners out to perform these tedious tasks. This garden work might then be redescribed as being a labor of love rather than the playing out of a macho male identity.

Gardening is a labor of love in two senses: (1) most gardeners enjoy working in their garden and get enormous satisfaction from the results of their work: a well-weeded border, a neatly raked gravel path or a clipped lawn; (2) gardening is a way to show affection and appreciation to others by helping them to pursue their goals.

Space, Control, and Social Image

Having an outdoor space to be in is an extremely important aspect of the garden to many. Partly this is about property and ownership. Gardening involves control of territory, time, and resources and how these will be deployed depends in part on different kinds of land claims, whether the garden is regarded as territory in the case of the private house or as tenure in the form of the rented house and garden or allotment (Ingold 1986:130). For most people their garden is the only private place outside their houses that they have the right to inhabit and control. Working in the garden is a way of establishing and maintaining the ownership of space. So it is not surprising that keeping this place as private as possible is a major concern for many. Hedges, walls, and fences exclude the outside, and the neighbors and their maintenance, are often major concerns in the English garden. The garden is a place in which one does not have to be sociable and get on with others. It may thus be deeply symbolic of personal freedom. Many gardeners feel that they have an absolute right to decorate their garden the way they like and most especially their back gardens that are concealed from the view of people passing by in the street. The exclusive control of the garden space is important in providing a sense of personal autonomy over decision making. The garden is a kind of kingdom over which the gardener reigns, and his or her subjects are the plants and the borders. The power of the gardener to control this place is absolute and this in effect empowers him or her in other ways that may not be possible.

Garden centers and nurseries are major sites for contemporary consumer consumption with the garden itself acting as an arena for individual display (Bhatti and Church 2001). In this respect it is interesting to note that many social democrat politicians were deeply suspicious of the manner in which allotment gardening was being promoted in the early twentieth century in Sweden as a way of improving the lot of the working classes. Instead of demonstrating on May Day and reproducing a revolutionary class consciousness, the working classes would be on their own individual allotments planting their seeds distracted from political activity (Ek 1979).

Some gardeners appreciate their gardens because they feel emotionally alienated from their homes and the garden is a place over which they feel that they can have some power. For some men in particular the garden becomes a refuge from their wife and the family, from domesticity and control, hence the popular image of the male of the household lurking in the garden shed and ordering his own little world within it. For many women, work in the garden allows them to escape from household chores, regarded as being intrinsically unsatisfying as mentioned above.

Gardens may be regarded as transitional places between the inside, the home, and outside public space. Most have front spaces that are relatively public and back space regions relatively free from the view of others. Status emulation, impressing others by the manner in which the garden is maintained in a neat and tidy manner, and the conspicuous consumption entailed by growing fashionable plants bought in garden centers is usually far greater a concern in the front of the garden than in the back, especially in the suburbs. It is quite clear that the degree to which the garden is maintained directly reflects on the social standing of the gardener. An unkempt and untidy garden thus becomes a reflection of personal and social morality. Someone who does not care adequately for their garden and allows the weeds to grow does not care about his or her neighbors and risks ostracism as a result (Grigson 2007:144). It is of little surprise that the most orderly and tidy of gardens in which everyday little acts such as picking falling rose petals off the drive because they look untidy may become a major concern in a situation in which everybody is watching everyone else. It is possible to entirely neglect the garden only in certain areas of the inner city or deep in the countryside where nobody cares.

Fame and Garden "Capital"

What a garden means and why it is important is intimately related to the types of garden people have and their knowledge of plants and techniques or how much "garden capital" they possess. I use this term, adopted from Bourdieu (1984), to refer to knowledge and skill in gardening (see note 1). For example, we can consider the exchange of plants. In general plants are good to give and good to receive. Most gardeners give plants away to others and to do so may be a valuable way of maintaining social networks. To whom would you give a plant? If it is a common or ordinary plant, then generalized or balanced reciprocity (Sahlins 1974) is the norm. On the other hand if it is an expensive or rare plant it is only likely to be possessed and cultivated by gardeners with considerable knowledge and skill in gardening. They may just cultivate this plant themselves and not give it away. The possession of this plant then becomes a marker of their own distinction within the gardening world and part of their inalienable wealth like a family heirloom that one does not give away (Weiner 1992). However, in some cases this may be problematic. Some species are very tender and difficult to cultivate and propagate. There is always the risk that the plant will die and be lost forever. One strategy taken by some gardeners is to give one or a number of these plants away to other gardeners. They will be gardeners who are known to be keen and skilled and unlikely to neglect the plant or let it die. Such gifts are a kind of long-term banking or insurance arrangement whereby if the plants in your garden should die one can always ask the person to whom you gave the plant for a bit of it back again. In a classic Maussian sense (1990) their garden in effect contains a part of you that is recoverable should the need arise.

Individual plants have very different meanings for gardeners according to culture and class. Certain plants will be described by those with high gardening capital as being vulgar and garish. Most bedding plants favored in public parks and in the brightly colored hanging baskets outside pubs come into this category, and this is partly because they require so little knowledge and skill to cultivate. Annuals tend to be disparaged as "common" in favor of perennials.

Some keen gardeners are plant collectors and their gardens become essentially containers to house their collections. These gardeners collect plants in the same manner as other people might collect thimbles, tea towels, or stamps. Often the aesthetics of the garden, whether it looks and feels pleasing, or creates the right kind of ambience or aura—an important consideration for many—will be entirely neglected in order to cultivate as many different types of plants as possible. Some are generalists cultivating a wide variety of different species. Others are specialists and in extreme cases growing only one species of plant such as heathers, lavenders, or bamboos.

Under the auspices of the Royal Horticultural Society they may apply for and be certified as holding a national collection of a particular species. These gardeners may feel a heavy burden of responsibility in keeping their plant collections going for the sake of the nation. Holding a national collection of course is a great source of pride and contributes considerably to their fame in the gardening world. Their gardens and collections will be regularly visited and admired by other elite gardeners. People possessing high garden capital are not necessarily wealthy or well educated in a conventional sense. In Bourdieu's (1984) terms they may possess quite low economic or educational capital. Specialist plant cultivation allows them to acquire distinction and fame that in other spheres of life are not possible. In this sense gardening compared with many other areas of life is relatively democratic, and success within the gardening world may transcend wealth, class, and privilege.

Time and Memory

One of the great pleasures of gardening is that it is a seasonal activity and the garden alters according to the time of the year. Different tasks are undertaken and different phenomena unfold at different times: pruning roses and the smoky bonfires associated with burning leaves in the autumn, planting seeds and trimming in the spring, the hoar frosts of the winter in which the plants are turned magically white, the vibrant plant growth and colors of the summer. Beyond seasonal change the garden alters on a daily basis according to the weather and the conditions of the light from early morning mists to the bright light of midday to the arrival of dusk and nightfall. These constant annual and diurnal cycles are a very important aspect of the meanings of gardens to many. The garden is never the same place twice. The first thing that most gardeners do when they return from having been away from home for any period of time is to go and walk around and see what has happened to their garden. The constancy of the domestic interior in which the picture will not have moved on the wall and the sofa will be in exactly the same state as it was left contrasts markedly with the sense of surprise that the garden may bring. A garden because it is a growing thing, unlike a domestic interior, is never finished. The garden may require a great deal of work and many gardeners look forward to finishing off their chores in the autumn and putting their garden to bed.

It was noticeable how differently gardeners spoke about their gardens when they were walking around them with me rather than sitting in their houses talking about them. Our conversations became less abstract and much more laden with memories. People would point to a plant and tell me that it was a gift from a relative, or that they had bought it in a market, or that they had grown it from a cutting taken from a public garden or park or that it was something that they had brought back from a holiday. Gardens, to many, are deep memory groves. This does not only apply to the plants in the garden but to everything else in it: garden sheds, greenhouses, paved areas, archers, fences, trellises. People can often remember when and why and how they were purchased or made and erected and they come to have deep biographical significance for them. Visiting a garden once provides a snapshot through time. It will never be exactly the same place again, and unlike the visitor, the gardener always views his or her garden through the prism of time: how it is now and how it used to be, and how in their dreams it might be in the future. The future is absolutely fundamental: gardening is a future-orientated practice in which there is always the potential to create something different and be surprised by the growth and the agency of the garden itself.

A Cultural Sense of Being English

To some the garden was an expression of English national identity. Gardening was a quintessential part of English life. England was very much regarded as the world center of gardening culture with other nations being to various degrees on the periphery. When I asked people whether they thought their garden was English, most (excepting those who had deliberately created foreign gardens such as a Japanese garden or a Mediterranean garden) replied in the affirmative. Many referred to the cottage garden as the style of garden that best expressed English identity: a rural garden with a profusion of different kinds of plants all mixed together, roses around the door, a little untidy, a bit improvised, nothing very formal, grand, or pretentious. Yet very few English gardeners actually live in the countryside and most had not actually attempted to create this type of garden in an urban or suburban setting. Although the cottage garden was a normative ideal of what a garden should look like, the majority of gardeners were quite happy with what they had that was surprisingly diverse and varied (Tilley 2008). It was in fact the making and maintaining of gardens that was most important to them rather than the end result. In fact it would be peculiarly non-English to conform to a cultural norm of what a garden should look like because that would in effect defeat the entire object of gardening that was individual expression. So it is the practice of gardening rather than the garden itself in which English national identity is thought to reside, being part and parcel of a long cultural tradition of gardening, that is, one of the practices of everyday life (de Certeau 1984).

Conclusion: Garden Voices

I have discussed above eleven of the principal reasons why people garden and what their garden means to them. Here I want to introduce a selection of voices. Much anthropology tends to be written in the abstract. Although supposedly based on conversations with and observations of others, their words tend to be drowned out in an interpretative gloss in which the only words and interpretations that appear to be significant are those of the anthropologist. So as a counterpoint to this here are the ways in which some English gardeners have described the significance of the garden and gardening in response to the questions: "What does your garden mean to you?" and "What do you like best about your garden?"

Five London Gardeners Talking

> It becomes a part of your well-being. You get the feeling of well-being in the garden, that therapeutic thing. You feel that I have done everything in the garden. Every single plant, the lawn, every blade of grass, every leaf, everything here is what I have done some time or another, planted or done something to... Every single plant is my doing and if I don't like it I can change it. I'm not subject to the whims and moods of my bosses at work. (Patrick)

> It's an expression of creativity. At times I feel it justifies my existence... It's about recognition to a degree. It's about needing to be recognized, valued or known... The end object for me as a gardener is to let people in. (John)

> I can imagine three impossible things before breakfast like the white queen but I wouldn't like to. I daresay I could fill my time one way or another but it would be a loss, a considerable loss... I love propagation because it involves taking something that seems unlikely like a cutting or a seed and looking after it and finding one has a complete plant. (Charles)

> It's mine. That's what I like about it... It's something I have nurtured and created. (Nigel)

I think it is a retreat really. Yes. Your space you can go and hide. I like people, don't misunderstand me but I also like somewhere where I can go and hide and be yourself and do what you like and take as much time as you like as well. I think it is a retreat. When I used to work in a nursery and came back to my own garden all the hassles and stress of the day would just disappear because it was so beautiful to look at. And I know one's planted it yourself but it's the plants doing it if you like. And it's just balm to the spirit in some way. (Shirley)

Five Rural Gardeners or Gardening Couples Talking

This is the space around my house which is mine. And only invited people can come into it. And it is controlled. I don't like the idea of controlling nature really. Yes I do. (Angela).

It's your little outdoor private world... It's just lovely to think that you can go out into your own little world which you have created and relax. Your own private little world, your empire. (Tony)

"Well it's been everything to me. I had my mother here with Alzheimer's for 5 years and I would have gone mad without it if I hadn't had the garden, and I mean Richard, he has left a big house in the New Forest, a big farmhouse. He knew that I would never leave the garden. So he moved here" (Julie).

Every part is important to me and every plant and bush is important to me. I have terrible migraine and I have been out watering. Despite that I have been out in the garden in the summer because I can't bear the thought of things dying. (Pam)

I think it is a sense of tranquility and peace. If you feel at all ruffled about anything and have a chance to get into the garden it calms you. It's only because we love it that it has that effect. It takes your mind off troubles. (Gillian)

Yes that's the same for me. It's the only area around the house one can escape from what one's troubles are inside the house. And it's fairly difficult to think of much else than the garden and the beauty that surrounds it. We're lucky here with the view but you appreciate your surroundings much more outside. (Peter)

I spend virtually all the free time I have in the garden. I don't mean necessarily working although a lot of the time I am working in it... It is my space and my relaxation. I feel very content and happy. It's just such a good space to be. I can't imagine being without a garden however small it was. (Susan)

Notes

1. Fifty-five English gardens were studied and interviews were conducted with sixty-five English gardeners, some of whom were couples maintaining the same garden, twenty-six (40 percent) were male gardeners and thirty-nine (60 percent) female gardeners. Their ages ranged from 27 to 84. Rural, urban, and suburban domestic gardens were studied across southern England from London in the east to Exeter in the west. Their

occupations (or former occupations if retired) of these gardeners were very wide ranging and included the following: nurse, prison officer, lorry driver, rubbish collector, factory worker, dentist, optician, management consultant, teacher, parks director, academic, computer programmer, salesman, farmer, picture therapist, electrician, telephone receptionist, architect, massage therapist, air force captain, army general, judge, ornamental ironsmith, journalist, sales assistant, seamstress, company director as well as unemployed persons on benefit and housewives and househusbands. Approximately 20 percent of the interviewed were elite gardeners. These were persons who possessed high "garden capital," a term I adapt from Bourdieu (1984), meaning gardeners with a high knowledge, interest, and expertise in gardening. Such gardeners were generally members of gardening societies and opened their gardens occasionally to members of the public for charity or for private viewing (in England under auspices of the National Gardens Scheme). The possession of high garden capital was crosscut by social and occupational status and other factors such as gender, economic capital (financial resources), and educational capital (formal educational qualifications). Thus someone with high garden capital might, or might not be, of higher or lower economic status, be male or female, relatively poor or wealthy, or have formal educational qualifications or not, such as a university degree. All interviews were conducted with gardeners in their own gardens or allotments. These consisted of well-prepared as well as impromptu questions and were recorded.

2. All names have been changed.

References

Anonymous. 2001. "Battle of the Sexes." *Amateur Gardening*, 28 (September Special Issue):21-48.
Bhatti, Mark and Andrew Church. 2001. "Cultivating Natures: Homes and Gardens in Late Modernity." *Sociology*, 35:348-365.
Bourdieu, Pierre. 1984. *Distinction: A Social Critique of the Judgement of Taste*. London: Routledge and Kegan Paul.
Certeau, Michel, de. 1984. *The Practice of Everyday Life*. Berkeley: University of California Press.
Chevalier, Sophie. 1998. "From Woollen Carpet to Grass Carpet." Pp. 42-72 in *Material Cultures: Why Some Things Matter*, edited by Daniel Miller. London: UCL Press.
Cooper, David. 2006. *A Philosophy of Gardens*. Oxford: Clarendon.
Ek, S. 1979. *Kolonins Sista Strid*. Göteborg: Etnologiska Föreningen i Västsverige 6.
Featherstone, Mike. 1995. *Undoing Culture: Globalization, Postmodernism and Identity*. London: Sage.
Foucault, Michel. 1987. *The Care of the Self*. London: Viking.
Francis, Marcus and Caroline Hester. 1995. *The Meanings of Gardens*. Cambridge, MA: MIT Press.
Gell, Alfred. 1998. *Art and Agency*. Oxford: Oxford University Press.
Grigson, Nicky. 2007. *Living with Things: Ridding, Accommodation, Dwelling*. Wantage: Sean Kingston.
Hitchins, Russell. 2006. "Expertise and Inability: Cultured Materials and the Reasons for Some Retreating Lawns in London." *Journal of Material Culture*, 11:364-381.
Hoskins, Janet. 1998. *Biographical Objects*. London: Routledge.

———. 2006. "Agency, Biography and Objects." Pp. 74-84 in *Handbook of Material Culture*, edited by Chris Tilley et al. London: Sage.

Ingold, Timothy. 1986. *The Appropriation of Nature: Essays on Human Ecology and Social Relations.* Manchester: Manchester University Press

Kaplan, R. and S. Kaplan. 1995. "Restorative Experience: The Healing Power of Nearby Nature." Pp. 238-243 in *The Meaning of Gardens*, edited by M. Francis and R. Hester. Cambridge, MA: MIT Press.

Kingsbury, N. 2005. "As the Garden so the Earth: The Politics of the 'Natural Garden.'" Pp. 100-110 in *Vista: The Culture and Politics of Gardens*, edited by T. Richardson and N. Kingsbury. London: Frances Lincoln.

Latour, Bruno. 2004. *Politics of Nature: How to Bring Sciences into Democracy.* Cambridge, MA: Harvard University Press

Lewis, P. 1995. "Gardening as a Healing Process." Pp. 244-251 in *The Meaning of Gardens*, edited by M. Francis and R. Hester. Cambridge, MA: MIT Press.

———. 2003. *Making Wildflower Gardens.* London: Frances Lincoln.

Mackay, Hugh (Ed.). 1997. *Consumption and Everyday Life.* London: Sage

Malinowski, Bronislaw. 1935. *Coral Gardens and Their Magic.* London: Allen and Unwin.

Mauss, Marcel. 1990. *The Gift.* London: Routledge.

McCracken, Grant. 1988. *Culture and Consumption.* Bloomington: Indiana University Press.

Miller, Daniel (Ed.). 2002. *Home Possessions.* Oxford: Berg.

Ollman, Bertell. 1971. *Alienation: Marx's Conception of Man in Capitalist Society.* Cambridge: Cambridge University Press.

Sahlins, Marshall. 1974. *Stone Age Economics.* London: Tavistock.

Strathern, Marylin. 1988. *The Gender of the Gift.* Berkeley: University of California Press.

Tilley, Chris. 1999. *Metaphor and Material Culture.* Oxford: Blackwell.

———. 2006a. "Objectification." Pp. 60-73 in *Handbook of Material Culture*, edited by Chris Tilley et al. London: Sage.

———. 2006b. "The Sensory Dimensions of Gardening." *Senses and Society*, 1:311-330.

———. 2008. "From the English Cottage Garden to the Swedish Allotment: Banal Nationalism and the Concept of the Garden." *Home Cultures*, 5:221-252.

Weiner, Annette. 1992. *Inalienable Possessions: The Paradox of Keeping While Giving.* Los Angeles: University of California Press.

13
Making It, Not Making It: Creating Music in Everyday Life

Bryce Merrill

> "I have no interest in 'making it' in music. Just making music is it for me."
> Jonathan, home recordist

Only a few decades ago musicians interested in recording music had little choice but to use professional studios, which meant paying recording experts and sound engineers for their time, equipment, and expertise. Because cost was prohibitive, most musicians relied on live performances to generate enough interest to convince record executives to take a chance on recording. Few musicians succeeded, and most left no record of their creative legacy.

The years since have seen tremendous innovations in recording technologies and novel usages of these devices to create recorded music. The most significant technological development is the advent of recording devices that are small, simple, and affordable enough for most musicians to use outside of professional studios. These technologies initiated a self-reliant ethics toward music production and distribution as well as antagonism toward the music industry. The technology afforded musicians the opportunity to make music without being in the business of making music. Alone and as individual composers, contemporary home recordists comprise a peculiar new "art world" (cf. Becker 1982) that requires special theoretical consideration.

Empirically and more broadly, the home recording movement provides a window for envisioning how individuals use new technologies to create meaningful lives and consequential social worlds. Unfortunately, there is little research on the subject and what does exist neglects to address how individual recordists meaningfully use recording equipment. Instead, much of the previous research (Jones 1990, 1992; Morton 2000; Taylor 2001; Theberge 1997) has illuminated the considerable role new music technologies have played in changing the mass production and consumption of popular music. Emerging technologies have enabled a unique musical phenomenon in home music recording, one marked by a shift from public to private productive spaces and made possible by affordable yet reasonably sophisticated recording equipment. Steven Jones (1992), for example, discusses how "demoing," or drafting songs on home equipment to be performed later in professional studios, represents a novel step in the process of recording popular music made possible by the domestication of recording technology. Enabled by technology, musicians explore and arrange songs at home, unfettered by the monetary and temporal constraints of professional studios.

Paul Theberge's (1997:5) research examines "the role of recent digital technologies in the production of popular music." Theberge argues that new digital technologies have transformed contemporary musicians into "consumers of technology" (6) and that this transformation has altered musical practices and cultures in important ways. In the case of home recording, the ubiquity of affordable home recording devices creates new sites of musical practice. Thus, for some, as Theberge argues, "the privacy of domestic space becomes the ideal site of musical expression...rather than more public realms of night club and stage" (218). Theberge and Jones reveal how innovations in home music recording equipment have altered creative and cultural dimensions of popular music. While these studies are important for understanding the effect of technologies on popular music, the insights are limited because of their focus on popular music.

Pinch and Bijsterveld (2004:635) suggest that in the growing field of Sound Studies, scholars have begun to address how sound matters to social life and its students. "Developments in sound technologies over the last fifty years," they write, "have dramatically changed the way that music is produced and consumed." Although studies of these changes and their effects, such as Tia DeNora's *Music in Everyday Life* (2000) and Paul Theberge's *Any Sound You Can Imagine: Making Music/Consuming Technology* (1997), among others, offer great insights into how technological developments have changed music production and consumption, the emphasis in these works remains either on the professional creation of music (e.g., in professional studios) or on listening to music in everyday life. However, it is important to note that, empirically, emerging technologies, such as home music recording equipment, have enabled types of music production accomplished by nonprofessionals in everyday life. We may know how making sounds in professional studios and listening to sounds in everyday life matter (see also Bull 2000), but we have little understanding of how making sound in mundane places also matters.

In this chapter, I address the importance of technologically enabled acts of music making in everyday life. I focus on how specific recording equipment emerged to create the material conditions for a popular form of home recording disconnected from pursuits of profit or audiences. Small, affordable, and easily usable recording devices have emerged to enable musicians to make, as Pierre Lemonnier (1993:2) refers to it, a "technological choice" to record music as a hobby. In this chapter, I detail three recording innovations that directly influenced recording as an everyday musical practice: analog recording on cassettes, recording on digital tape, and computer-based recording. These developments, along with innovations in microphones and synthesizers (Jones 1992), have laid the material foundation for home recording as a locally important musical practice. I also take an interpretive approach to this phenomenon, focusing ethnographically on home recording practices and accounts of recordists who make music in their homes and for themselves. My aim is to present home recording as a unique, technologically enabled musical practice that occurs in domestic spaces but affects personal ones.

Method

In 2004 I began collecting data on home recording using ethnographic and autoethnographic methods in the Front Range area of Colorado. I classify all but three of the participants to this study as "hobbyist" recorders, insofar as participation in home recording is for reasons other than public acclaim. The three who have public aspirations (e.g., success in the music business, sought out attention by outside audiences) also record at home for recreational reasons, in addition to professional ones. I focus on the identity of these participants as home recordists as it relates to their musical practices.

I began my research by asking home recordists about their practices and motives. Data from these informal, unstructured interviews were used to create a semistructured interview schedule for additional participants. For example, one acquaintance discussed his practice of cataloging songs, which included keeping notes on technical and personal details of recording. I translated this information into a series of questions related to the practice of home recording. This line of inquiry ultimately produced the recurrent theme of "memory" addressed in this chapter. I derived analytically the thematic categories "catharsis" and "creativity" in similar fashion.

I recorded and transcribed thirty of the interviews, and took extensive field notes when participants either refused tape recording of interviews or when recording was not possible. In eleven cases, I took extensive notes during and after conversations and transcribed them immediately following the interview. Interviews lasted between one and a half hours and four hours.

I included in this dataset reflexive observations of my own recording experiences. During my research I recorded twenty-five songs, with each recording taking, on average, 5 hours. I wrote memos while recording, documenting the process, choosing the song topic, and writing down the personal response. I also took notes on my cataloging practices. I use these data to compare my experiences and narratives with others, and I analyze them only in concert with other data. The aim was to document my home recording experiences in connection with other recordists. The result of this research is a rich empirical look at the local, meaningful practice of home recording, one that is uniquely available through ethnographic inquiry.

Three social psychological themes resulted from my analysis. Home recordists use recording technologies to experience creative self-expression and exploration; they record to create, revisit, and revise memories; and they make recorded music to experience catharsis in their everyday lives.[1] These themes of creativity, memory, and catharsis reveal analytically the personal benefits recordists derive from using home recording equipment. Everyday actors employ recording techniques and technics to create their music and themselves.

Technics, Techniques, and Technicians

Chris Tilley (2001:260; see also Vannini chapter one) writes that "[o]ne of the most influential theoretical perspectives influencing material culture studies has been an emphasis on objectification: that through making things people make themselves in the process." Moreover, Tilley suggests that technologies "may be attributed agency...because they produce effects on persons" (260, original emphasis; see also Kien this volume; Vannini chapter five). Therefore, ethnographies of material culture such as this one should explore technologies as objects and as agents. Technology's agency emerges from the interactions of user and object; it is located in the triad of technic, technique, and technician.

I rely on René Lysoff and Leslie Gay's (2003) concepts of the ontological, pragmatic, and phenomenological dimensions of technology to consider the material properties of specific home recording technics, the practices (or techniques) of home recordists engaged with these technics, and the personal consequences of these practices. The ontological status of a technic refers to its physical properties and capabilities, and the second dimension—the pragmatic—refers to the uses, practices, and "forms of knowledge" associated with a particular technology (Lysoff and Gay 2003:6). While the ontological status of a musical technic, such as a piece of recording gear, is an object of specific capacities, the pragmatic dimension of this technology is dynamic, depending on its uses, explicit or otherwise, and the meanings associated with its use. For example, an analog portable multitrack recording unit is objectively a device that allows users to record sounds onto tape in particular ways. However, its uses (by amateurs or professionals), reasons for usage (to demo songs or to get an "analog" sound), and definitional status (helpful tool or sonic manipulator) are contingent upon people and their techniques, not the ontological status of the technology. Finally, the phenomenological dimension examines how technics shape human experiences in ways not necessarily connected to their intended function (7). The intended function of an analog recorder may be to capture and reproduce sound, but by creating recorded music with this device, recordists may experience a sense of catharsis. The machines are not built to become mechanical therapists, but people use them this way.

The intended function of home recording technics is to record music; however, recording techniques produce memories, catharsis, and expressive enjoyment. These technics function to produce people—technicians. Pierre Lemonnier (1993:2) suggests that social studies of technology must be aware of the obvious or intended functions of technics. Because, as he writes, "techniques are first and foremost social productions" (3), and we cannot assume that people will use technics as they are intended or most effective. Instead, we must investigate techniques as practices that are "simultaneously embedded in and partly a result of non-technical considerations" (4). This does not mean we should ignore the technical in favor of the social, but that we pay equal consideration to what things are and do and how we use them. Lemonnier's warning, shared by others (DeNora 2000; Latour 1991), is that we obfuscate such insights when we overlook nontechnical dimensions of technology and society. Instead, we must explore, as Streeck (1996) suggests, how people do things with things.

From Edison to American Idol

I want to explore the ontological statuses and histories of three important "things" to home recordists: analog, digital, and computer-based recording technologies.[2] Their historical development, like that of most technologies, is not direct or predictable (cf. Pinch this volume). Sharing a past with Edison and other pioneers of sound recording, as well as Nazi scientists, among others, the legacies and paths of development of these technologies are complicated, and it is impossible, without risking meaningless regression, to locate distinct beginnings (Chanan 1995). It is equally daunting to catalog all of the technological innovations that in some ways contributed to the emergence of home recording technics. Jones (1992:141) documents additional innovations and influences that contributed to the development of contemporary home recording devices, such as the emergence of the integrated circuit that helped to reduce the sizes of these technologies. Here I have foregone a comprehensive account of the development of home recording technics to focus on three integral developments. These accounts of innovation provide historical context and demonstrate that the future of technologies often lies in how everyday actors use them.

Home music recording as a large-scale endeavor evolved in part through indirect means and partly with strict purpose. Those with expertise and access to home recording equipment, famously Les Paul, began much earlier than the masses. Widespread recording began when consumers used the limited recording functions on home tape players. "Boomboxes," as they were colloquially known, were not marketed as home recording devices. They were often equipped with microphone inputs, and users could alter tapes to enable recording by tampering with write-protection devices. By the late 1970s, portable tape players like the Sanyo M9998 model were equipped with external microphones for recording. Internal and external microphones allowed consumers to record on their own. However, these crude means obviously resulted in the roughest of music recordings, products that could not compete sonically or commercially with even the earliest forms of recorded music.

Boomboxes were also extremely limited in their ability to record multiple tracks. Using two players simultaneously, one could record the first track on one player and additional tracks onto another player while playing along with the first track. This basic and inefficient form of multitracking paled in comparison to multitracking processes already in place in professional studios. Arguably, however, it functioned as a foundation for later home recording devices, which would ultimately rival professional technologies.

The American division of the Japanese conglomerate TEAC introduced the first dedicated multitracking home music recorder, the TASCAM 144 Portastudio. TASCAM was previously in the business of professional-grade recording equipment, particularly multitrack recorders and sound mixers. No recording and mixing device for home use were available until the Portastudio's introduction in 1979. Of course, the Portastudio was not the first home recording device—reel-to-reel recorders predated it—but it was the first to use cassette tapes, an incredibly affordable recording medium. This device also contained a revolutionary feature invented by TEAC, which allowed musicians to record additional tracks while listening to previously recorded ones. This Simul-Sync feature allowed home recorders to record much in the way studio engineers did. The sonic quality and diversity of applications on the Portastudio did not rival studio gear, but at least the option to record music without a professional studio was possible. Moreover, the US$1,100 retail price of the Portastudio was a pittance compared with the considerable cost of studio time.

Home recording in the 1980s was almost exclusively the cassette tape's domain. Throughout the 1980s, TASCAM continued to produce a line of affordable (under US$3,000) tape recorders for the home enthusiast. These models were similar to their predecessors but far more sophisticated. With their MINISTUDIO line, they introduced basic, battery-powered recorders sold for under US$350. The cost, size, and ease of use of these recorders encouraged more hobbyists and musicians with small budgets to record, but these machines were well behind the technology available to professionals, where digital recording was gradually overthrowing the reign of tapes.

The digital revolution began in the telephone industry and ran through military and scientific endeavors before making an impact on recorded music. Thomas Stockham, a professor at the University of Utah, made the first digital music recording in 1976. Stockham previously experimented with digital sound recording while at MIT, and formed the first digital recording company in America, Soundstream. Stockham's first digital music recording was of the Santa Fe Opera. Later, a rival company, Telarc, recorded the first digital music for commercial sale. Subsequent digital recording devices developed by different companies and in different parts of the world retained the basic premises of digital audio recording. They also shared three additional important similarities: cost, size, and complicated sophistication. It was common for these recorders to cost upward of US$30,000. Some models, such as the Australian Fairlight Series III, sold for over US$100,000. Digital recorders were also designed by and for knowledgeable sound technicians and engineers, not for the average person. Thus, their use was solely in professional studios. Finally, these early digital recorders were furniture-sized pieces of gear, taking up entire walls of studios. They were hardly fit for home use, nor were they meant for it.

Digital music recording technology ultimately made its way into home studios in the mid-1980s with the introduction of Alesis Digital Audio Tape (ADAT). Sony/Phillips developed Digital Audio Tape (DAT) as a recordable version of the compact disc. Using available analog-to-digital conversion processes, the DAT recorders were popular among recording professionals but not home recordists. Both professionals and hobbyists used 8-track ADAT recorders developed by the Alesis electronic music company. These were small enough for home use, affordable, and sacrificed very little in the way of sonic quality. ADAT recorders marked the digital revolution's transformation of home recording, allowing hobbyists to come closer than ever before at making studio-quality recordings. As innovative as this was, another sonic revolution soon made it obsolete.

Through the late 1980s and mid-1990s, Minidisc recorders, the more successful relative of Minidisc players, and cassette recorders were more prevalent in hobbyists' home studios, but ADAT recorders also held a share of the market. All three took a distant backseat to computer-based recording, a new recording technique with roots also in mid-century sound recording and reproduction technics and their developments in science, military, and industry.

In the late 1950s, a scientist at Bell Telephone Labs, Max Mathews, wrote one of the first music recording programs, a software program called MUSIC. The software, run on an enormous and powerful computer, underwent several innovations and sparked considerable scientific interest in computer music and sound recording. Universities throughout the United States and Europe, as well as companies like RCA, worked to improve this new recording format. Academic societies like the Audio Engineering Society (AES) devoted special attention to this new frontier of recording, but most of it was entirely scholastic. The early days of computer recording were shaded from the gaze of the general public—specifically home recordists—largely because computers needed to run these programs were not available to the public. They were large, expensive, and remotely user-friendly. However, as computers changed in size, expense, and ease of use, so did the dynamics of computer-based recording.

The Digidesign company is credited with offering publicly the first computer-based recording program "Sound Tools" for Apple's Macintosh computer. This program used advanced audio analog-to-digital converter, which transforms analog audio signals to numerical values (digital). It was also the predecessor to the more popular recording program, ProTools. Introduced in 1991 as a music workstation, ProTools provided professional studios with integrated recording hardware and software. It quickly became the industry standard for sound recording in film, television, and music.

Subsequently, Digidesign released ProTools models specifically designed for home use. Several other companies followed suit and released affordable computer-based recording programs for home use, such as Cakewalk, Reason, Cubase, and many others. Though somewhat different in insignificant ways, these programs offer home recorders the ability to record, engineer, and produce professional quality music in their own homes. Even more remarkable is a program like Apple's recording software GARAGEBAND, an extraordinarily basic and user-friendly recording program that comes free with Apple's basic entertainment package I-Life. Granted, the computer is not for free, but the program is. The products of a century of scientific innovations and innumerable financial costs paid largely by corporations and governments came home to roost in small, remarkably simply home recording programs.

Retailing for less than fifty dollars, the American Idol Extreme Music Recorder software allows aspiring singers record their own voices over music and remix existing songs, preparing them, perhaps, for a chance to impress

Idol's Paula Abdul or be belittled by Simon Cowell. It may be true that the amazing technological evolution leads to products like this one, but it has also led elsewhere, far beyond the glitz and glamour of the music business. Without doubt, it has moved out of the spotlight and into the homes of a new group of musical practitioners: home recordists.

Making Music...and People

> "The technology is available now to anyone even half interested in recording at home."
> Peter, home recordist

> "I am deeply, personally moved to do this."
> Taylor

William Gaver (1991) uses Gibson's (1979) concept of technological "affordances" to emphasize thinking about technics as devices that make certain outcomes possible, but these possibilities are realized only through the interaction of user and object. "An affordance of an object," Gaver (79-80) writes, "refers to attributes of both the object and the actor. This makes the concept a powerful one for thinking about technologies because it focuses on the interaction between technologies and the people who will use them." New recording technologies obviously afford individuals opportunities to make music at home, as Peter, quoted above, affirms. The uses of these technologies, however, are neither always nor necessarily related to their intended functions. In this section, I demonstrate that recordists use these devices not just to create music but also to create themselves. In other words, the affordances of recording technics are as much psychological in nature as they are musical.

The home recordists in this study share the belief expressed by Taylor quoted above. They agree with some variations on the theme that they benefit deeply and personally from the availability of home recording technology. Recording satisfies their intellectual and emotional need to create. Cognitive and emotional compulsions to record suggest that these home recordists value aesthetic considerations over technical or productive ones. Gary Alan Fine (1996:178) discusses the term "aesthetics" to capture the "'cognitive' ("satisfaction") and affective ("sensory") components of aesthetic judgments" and to "include the intentional qualities of human action." The chefs in Fine's study of professional kitchens make aesthetic evaluations and choices in their work that are distinct from other technical or professional considerations (178). Decisions to make something—a song or an entrée—because it is beautiful or feels rewarding are often as important as decisions guided by functionality.

The problem, as Fine puts it, for the "slippery" term "aesthetic" is that evaluations like "beautiful" or "joyful" are rarely defined ethnographically. In other words, critics and experts define aesthetic qualities, instead of everyday actors who undoubtedly employ aesthetic judgments in everyday life. Perhaps mundane life is more drab than beautiful, and everyday people do not have the training to judge beauty. A more likely explanation is that common approaches to studying aesthetics neglect the omnipresence of aesthetic judgments and experiences in social life (Vannini and Waskul 2006:13). Vannini and Waskul propose an aesthetic approach to studying everyday life using Dewey's (1959[1934]) work on art as experience. Their project is to develop the metaphor of "symbolic interaction as music," which is not directly related to this project, but their assertion, inspired by Dewey (Vannini and Waskul 2006:14), that the "esthetic potentially informs all kinds of experiences and meanings" is important. It is relevant because recordists in this study express the multidimensional aesthetic values of home recording; aesthetics are important ethnographically.

Home recordists reveal that aesthetic experiences indeed happen vibrantly and consequentially outside of formal art worlds. In their everyday lives and in ordinary places, home recordists create for the sheer pleasure of creation. The inspiration to record is guided by aesthetic concerns—pleasure, joy, beauty, reconciliation, preservation, and so on—such that music is often a secondary concern. Their technologies, in fact, allow them to erase recorded songs in an instance, leaving no material record of their actions, but this does not mean that deleted songs are not productive. They are productive insofar as the recordist experienced artistic creation. To borrow Vannini and Waskul's (2006:14) adaptation of Dewey's "art as experience," the art of home recording is experience.

Unlike professional studio recording, home recordists face fewer and lesser rigid time and scheduling constraints. Most take full advantage of this by recording whenever they have spare time. Some set aside specific times to record, but most do it whenever they feel like. William, a home recordist with a studio in a spare room, confesses:

> I try and set aside time to record, I really do, but it rarely works out that I stick with a schedule... I usually kind of sneak away to record when I get the urge...after work for an hour or during work if I'm working at home!

As someone who also works at home often, I identify with William's approach. The urge to create music often comes at the unexpected hour—such as while writing or grading papers—but it is an easy urge to satisfy when the means to do so are so readily available. In this way, the creative process takes on a dynamicity otherwise not allowed with conventional recording practices. This is a particularly important aspect of home recording to Emily, a recordist who records a lot both in professional and home studios. She glows as she says:

> I'll be in the shower, and I'll hear some rhythm...maybe from the water or a cat that has joined me in the bathroom, and I'll start tapping it out on the tile, and I'll think, "This is great. I don't want to lose this." My recording gear is in a big closet in the bedroom, where the bathroom is, so I'll get out [of] the shower—dripping wet sometimes!—and record the beat. It's so great that way.

What is "so great" for Emily and many others is that inspiration and recorded performance often enjoy the cozy comforts of home recording. That is, home recording integrates more closely, if not more conveniently, the inspiration to record and the act of recording.

This lack of the typical time constraints of recording also changes how recordists go about creating music. Feeling no pressure to finish composing by a certain deadline, they often work on songs for only as long as they desire. Sometimes this means finishing a song in a single session; other times it means working on something over several sessions. Interestingly, it also means that songs often go unfinished, insofar as a final product is never realized. However, as one participant aptly put it, finished products are rarely the point:

> I find myself not finishing songs, writing a lot of parts, you know? This might not be good for me as a songwriter, but it doesn't bother me that much. I mean, when I'm finished with something—a full song, a scrap of a song, or just some rhythm—I'm finished with it! That's all I wanted to do. That's all I have to do!

What Larry is suggesting is that his creative urge is satisfied regardless of the realization of a musical product. Considering a potential contradiction here, I asked Larry why he would bother to record some bit of music if he was not going to complete it:

> B: Why record it? Why not just play it?
> L: Because I might just want to remember it, because I might be trying to work out other parts to it [pause] and because it's fun to record stuff.

Larry's point is that home recording is useful regardless of the outcome. The practice itself is both functional as a tool for writing and enjoyable as a means for self-fulfillment. In this way, home music recording is more akin to gardening, stamp collecting, or diary writing—aesthetically driven practices with few tangible outcomes—than it is a strict form of music recording.

I do not wish to overstate this parallel between home recording and other hobbies. They are similar in their emphasis on personal enjoyment; music is more central to these recordists, many of whom are or were practicing musicians in other ways. Jonathan, a well-off certified public accountant, made this point clear. He began his musical career as trombone player in an army brass band and toured extensively. Music was his career for a long time, but he left it when he wanted to start a family and provide for them beyond what his musician's salary would afford. He was eventually able to afford a home studio, which he uses to make music with his children and wife and keep music in his life, if only in the margins of it:

> I love playing music, always have. But here's the thing I always come back to: I had to become an accountant to make a good living; now I can afford this great home studio and play music, my real passion...but my work and family, which takes a lot of time, still come first...so I can't play music all of the time like I used to, but with this studio, I can still have music in my life...and now I can share it with my kids and my wife.

For Jonathan, making music at home is not the same as being a professional musician, but it is still making music. That is what is important to him and to many others who record at home.

Every participant I interviewed expressed a genuine love for making music, but most struggled and some disagreed vehemently with my equating home recordists with the social identity of a "musician." "I'm not sure if what I do counts as being a 'musician,'" Julius told me. He went on:

> To me, being a musician means all of these other things, like playing live and selling music. What I'm doing is recording music in my basement...for no one, for no good reason. Is that being a musician? I don't know.

This distancing of himself with this particular category reinforces the idea that home music recording differs in important ways from other musical endeavors. So much so that Julius identifies himself with being a "home recordist" rather than a musician. His practice and that of others are qualitatively different from those typically associated with working musicians. Home recordists rarely, if ever, perform their material live, and their audiences are few, if anyone. What is important to the home recordist is the practice of making music.

A question that drove my initial inquiry into home music recording is why make recorded music for no one? The answer to this question, as I have found, suggests that the question itself is problematic. The answer is, as Travis put it, "I do this for me." Their music is not made for no one; it is made for the home recordist. I revised my question: "Why make music for yourself?" and it has a much more intriguing answer, which speaks to the phenomenological dimensions of material culture (see Richardson and Third this volume) and in particular home music recording. Practitioners of home recording value the importance of making a sonorous, material product, and all that these practices and experiences offer to themselves. In social psychological terms, I want to address three ways that home recordists express these experiences: the mnemonic, creative, and therapeutic dimensions of home recording. Each category reflects an emergent theme in my ethnographic data and suggests that experiencing music in these ways is the primary motivation behind home music recording.

Home recordists, as Larry mentioned previously, use recording as a tool for remembering musical parts or arrangements. However, they use home recording for narrativization (see Woodward this volume) to remember people and places and events and many other natural subjects of memory. Some, like Brett, make sophisticated catalogs of their songs with the memories they reference. Brett's catalog contained over thirty songs that he had given titles to and added brief descriptions of what (or who or where) the songs were about. He returns to his songs and catalog periodically to be reminded of whatever particular memory to which he wants to listen. Others like Larry have no such systematic methods, but value their collections in the same way:

> Every now and then I go back and listen to some shit I did in high school or something and think, "Oh, man, that was stupid." But it's pretty cool just to go back and think of recording it then and what was going on then.

Larry's statement also suggests that not only does he remember what he recorded and why, but when he recorded it. Mark made a similar comment when he told me that listening to previously recorded music brings up "all sorts of memories, including how bad I was at recording back then!" In each way, home recording technics provide the material conditions for engaging in mnemonic techniques. These techniques also reveal the creation of types of technicians—the capable or incapable technician is but one type of a technician created.

The mnemonic and therapeutic dimensions of home recording are connected. Practitioners often use home recording to process or resolve particularly painful events, often surrounding the common musical theme of heartbreak and love lost. Recordists like Michael often find their productivity skyrocketing during romantic turmoil and plummeting in romantic bliss. With a slanted smile, Michael tells me:

> I'd write these songs after breaking up with a girl...later after I was happy because I was seeing somebody again, like now, I'd listen to these songs and think, "Man, was I pathetic!" So I'd either delete them or rework them, usually change the lyrics, so they weren't so depressing.

In this example, Michael relies on home recording as a form of objectification capable of resolving his troubles. In the therapeutic process, he also creates a material memory of some point in his life, which he can now reflect upon and, in this case, revise his feelings. The therapeutic and mnemonic dimensions of home music recording are not always so redemptive, however. Julius is a home recordist who experienced at a young age the tragic loss of his father. He has written and recorded a song about it, but he is less than optimistic when I ask him if this has helped. He replies, "I don't know if it's helped. I mean, it's there, but I don't know if it's helped." The record of the song is there, but this particular memory apparently holds little therapeutic value.

The experience of being creative seems to be the strongest motivation for home recordists. Certainly, this holds for most musicians, but all participants express the creativity derived from home recording as being of a kind that is liberated from the normative constraints of music and, separately, the hassles of the music business. Consider this comment by Murray:

> I'm no musician, man, but it doesn't really matter. When I record music I can do whatever I want...fuck up as many times as I want...and all that matters is doing it. I don't have to worry about someone telling me I'm off beat or in the wrong key. I'm doing it!

Murray's liberated creativity is in no small part due to the editing functions of recording equipment. Computer, digital, and analog recorders enable repeated performances and playbacks. Mistakes, which would be costly in professional studios, are yet another part of the creative process. In fact, all of these personal benefits of recording at home are inextricably linked to the technical capabilities of recording technics. Again, this story of home recording cannot be told without equal consideration of technics, techniques, and technicians.

Wesley creates music that may typically cross genres in unacceptable ways, but that is not his concern. "I can play funk parts and classical parts at the same time, if I want to, because that's what I want to do. Doesn't really matter if it's supposed to work," he tells me while laughing. Every participant who was or is involved with music delights over the practice of recording music without the pressure, real or imagined, of taking into account what commercial audiences will think of it. Ben, an accomplished musician, remarked that the lack of pressure with home music recording enhances his experience. He still feels creative in professional studios, but on his own, Ben tells me, "I'm at home with music. You know? I'm at home with it." Being at home with music, in the literal and figurative sense, is the ultimate appeal for home recordists like Ben.

Home for Music

Home recording is technocultural practice that is enabled by material objects, shaped by recording practices, and consequential, personally, for its practitioners. It is a unique form of musical creation that cannot be understood solely in terms of its impact on popular music. Instead, I have attempted to demonstrate that the story of home recording cannot be told without serious consideration of how emerging technics are used by people in their everyday lives. People now record music at home not for public consumption, not for monetary gain, and not for acclaim. They do it because the practice is aesthetically rewarding. They do it to remember; they do it to heal; they do it because it is fun and they do it for themselves. Recording music enriches the self, if not the landscape of popular music.

The question that remains is not "why do people do it," rather "what effect does this have on the rest of us?" I am not convinced that the insular relationship home recordists have with the music industry means that they have no effect on music or society more broadly. I do know that their effect is unusual and, to date, not considered by scholars of culture, music, technology, or social change more broadly. What is typically considered is how the products of new media technologies influence the typical social and cultural outlets (e.g., how blogs change journalism, digital video cameras change the film industry, home recording devices change the music business), and in doing so we miss a critical dimension of these technologies and their place in society. That is the way that these technologies have changed people in their everyday lives. Indeed, the "emancipatory moments" of these technologies, to use Mathew Malsky's (2003:254) phrase, may quickly disintegrate through corporate co-optation, but their liberating potential appears elsewhere. In the case of home recording, it can be found enriching and changing in important ways the lives of people practicing music in its newfound domestic margins. It can be found subverting the rules of normative music construction, detaching creativity from fame, and playfully, if not always happily, helping people to cope with the trials of existence. In all ways, home music recording affects people; people inevitably affect the social worlds in which they live.

What is needed with this and other similar subjects (like evolution and usage of personal digital and video cameras and Web-based diaries or blogs) are insights into how new media technologies are changing and challenging conventional forms of artistic practice and, thus, are changing cultural and social landscapes of contemporary societies. Additional analytical investigation into the practice of home music recording should reveal how so many people staying home for music affect society.

Notes

1. Using Robert Perinbanayagam's (2006) concept of reflexive catharsis, I view catharsis as a temporary and situational state of pleasure brought on by emotional release.
2. I have provided several links to Web sites with historical data on the social history of this technology in the "references" section. Those who are interested can read about developments discussed here and more at these Web locations.

References

Becker, Howard S. 1982. *Art Worlds*. Berkeley: University of California Press.
Bull, Michael. 2000. *Sounding Out the City: Personal Stereos and the Management of Everyday Life*. Oxford and New York: Berg.
Chanan, Michael. 1995. *Repeated Takes*. New York: Verso.
DeNora, Tia. 2000. *Music in Everyday Life*. Cambridge; New York: Cambridge University Press.
Dewey, John. [1934] 1959. *Art as Experience*. New York: Capricorn.
Fine, Gary Alan. 1996. *Kitchens: The Culture of Restaurant Work*. Berkeley: University of California Press.
Gaver, William W. 1991. "Technology Affordances." Pp. 79-84 in *Proceedings of the ACM CHI91 Human Factors in Computing Systems Conference*, edited by Scott Robertson, Gary Olson, and Judith S. Olson. New Orleans, Louisiana.
Gibson, James J. 1979. *The Ecological Approach to Visual Perception*. Boston, MA: Houghton Mifflin.
Jones, Steve. 1990. "The Cassette Underground." The Cassette Mythos: Autonomedia. Accessed November 1, 2006. Available at: http://cassettemythos.com.
———. 1992. *Rock Formation: Music, Technology, and Mass Communication*. Newbury Park, CA: Sage.
Latour, Bruno. 1991. "Where Are the Missing Masses? A Sociology of a Few Mundane Artefacts." Pp. 225-258 in *Shaping Technology/Building Society: Studies in Sociotechnical Change*, edited by Wiebe Bijker and John Law. Cambridge, MA: MIT Press.
Lemonnier, Pierre. 1993. "Introduction." Pp. 1-35 in *Technological Choices: Transformations in Material Cultures since the Neolithic*, edited by Pierre Lemonnier. New York: Routledge.
Lysoff, René and Leslie Gay. 2003. "Introduction: Ethnomusicology in the Twenty-First Century." Pp. 1-22 in *Music and Technoculture*, edited by René T.A. Lysoff and Leslie Gay. Middletown, CT: Wesleyan University Press.
Malsky, Matthew. 2003. "Stretched from Manhattan's Back Alley to MOMA: A Social History of Magnetic Tape Recording." Pp. 233-263 in *Music and Technoculture*, edited by Rene T.A. Lysoff and Leslie Gay. Middletown, CT: Wesleyan University Press.
Morton, David. 2000. *Off the Record: The Technology and Culture of Sound Recording in America*. New York: Rutgers.
Perinbanayagam, Robert. 2006. *Games and Sport in Everyday Life: Dialogues and Narratives of*

the Self. Boulder, CO: Paradigm.
Pinch, Trevor and Karin Bijsterveld. 2004. "Social Studies: New Technologies and Music." *Social Studies of Science*, 34:635-648.
Streeck, Jurgen. 1996. "How to do Things with Things." *Human Studies*, 19:365-384.
"TASCAM Company History." n.d. Accessed October 1, 2006. Available at: http://www.tascam.com/Company.html
Taylor, Timothy. 2001. *Strange Sounds: Music, Technology, and Culture.* New York: Routledge.
Theberge, Paul. 1997. *Any Sound You Can Imagine: Making Music/Consuming Technology.* Hanover, NE: Wesleyan University Press.
Tilley, Chris. 2001. "Ethnography and Material Culture." Pp. 258-272 in *Handbook of Ethnography*, edited by Paul Atkinson, Amanda Coffey, Sara Delamont, John Lofland, and Lyn Lofland. Thousand Oaks, CA: Sage.
Vannini, Phillip and Dennis Waskul. 2006. "Symbolic Interaction as Music: The Esthetic Constitution of Meaning, Self, and Society." *Symbolic Interaction,* 29:5-18.

14
The Death of the Clinic: Domestic Medical Sensoring

Patrick Laviolette

> The marvels of modern technology fill us with amazement but also with dread. All the time we are haunted with nagging anxiety. Isn't the gadgetry getting too clever?... If the computers take over, where do the human beings come in at all? (Leach 1968:16)

So began the second session of Edmund Leach's 1967 Reith Lectures, "A Runaway World?" This opening to a 40-year-old broadcast still zealously captures the idea that our fascination with technological development in a mechanical age is riddled with paradox. This chapter examines such paradoxes through a case study focusing on domestic heart disease rehabilitation located within the contemporary British health system.

In the UK, over a million people a year are repeatedly admitted to hospital emergency wards in situations where it is argued that the right primary care managed outside hospitals could have avoided such drastic and costly measures (Fisk 2003). Shifting European demographics lead pundits to predict that one in seven residents will be over 70 by 2030. The UK presently has the highest proportion of residents over the age of 65 in Europe (18.7 percent). It is therefore likely that there will be a rapid increase in age-related medical conditions in Europe. Although the decreasing prevalence of disabling and lethal infectious diseases marks the triumph of Western public health care over the past centuries, such diseases are being replaced by complicated long-standing chronic illness. These are associated with extended longevity as well as relatively high social hygiene and affluence. They fall into three groups (systemic degeneration, biomechanical, psychological). The first—organic degeneration or systems failure—includes most cardiovascular diseases and these have been identified as the most pervasive and preventable in terms of residential care.

Consequently, the Department of Health (2000) has invested up to £80 million for providing domestic assistive technology in the form of telecare to all who need it by 2010. As a remote form of therapeutic rehabilitation from chronic conditions, telecare consists largely of two technological forms that work in collaboration with the prescription of medicines as well as architectural design. These two main mechanical technologies include health variable verification devices and lifestyle monitoring sensors (**LMS**). Health variables are examined with blood pressure and blood sugar gauges, scales to record weight gain/water retention, and an interactive personal "Docobot" computer that asks questions daily about one's health and sends the responses to a central database. Lifestyle monitoring (also known as lifestyle management/reassurance) consists of real-time Passive Infra-Red (PIR) motion sensors, appliance use detectors, emergency response alarms and such. The idea is to draw up patterns of healthy behavior to contrast with illness indicators that can trigger immediate responses from carers or specialists (Porteus and Brownsell 2000).

Telecare is therefore intrinsically dependent upon the principles and technologies of surveillance and monitoring. The comprehensive implementation of such devices and the interpretation of the data they generate are still, however, ill understood. Compelling justifications for championing this type of telemedicine is to save time, bed-space, and money for health services as well as allowing outpatients a greater awareness of their own condition and a level of independence from institutional care (Dant and Gully 1994). The inertia behind domestic care schemes that rely on various forms of information and communication technology is thus shifting the biomedical field from the hospital to the home.

But if such utilitarian and emancipatory ideals are to be balanced realistically, there must first exist comprehensive knowledge of how they work in site-specific contexts. At present, they are still under trial and tested as if to yield universal models of applicability. Indeed, the standard perception of Western science as an undertaking that exists beyond specific spatial particularities has stifled many of our epistemological understandings and practical social initiatives. Influenced by the work of Latour (1987) among others, Livingstone (2003) challenges those conventions that view scientific knowledge and the development of technological innovations as isolated from their localized influences. By providing historical sketches for certain places of scientific experimentation, he reveals the intimate relationships between sites of knowing and scientific truth claims. For their part, Law and Mol (2001:619) question the regionalized networks of technoscience and provide alternatives that "turn universality inside out" and that answer to a complexity of social relations. Here the immutable mobility of discovery is local, regional, and even global at the same time, yet at different moments, that is, in other temporalities. Picking up on seminal themes that date back to Heidegger (1971 [1952]), Kuhn (1962), Wittgenstein (1969), and Feyerabend (1978), we find that these scholars have been among the principal advocates for the recent contextual and spatial change in the discourses of science and technology studies.

This chapter, based on my participation in a large research consortium pilot implementation, attempts to ethnographically contextualize this technological package of therapeutic material culture. In addressing the issues of aging, chronic disease, and domestic care, it also fits into wider medical anthropology concerns for exploring the spatial materialities of science and technology developments because ultimately telecare is the provision of health care at a distance, with the support of information and communication technologies. By examining contemporary everyday experiences of telecare health monitoring for Chronic Heart Failure (CHF) sufferers over 60, living in South Yorkshire, UK, this chapter explores some of the spatially materialized paradoxes implicated in the provision of domestic assistive technology.

STS: Scapes of Telehealth Surveillance

In thinking about how different kinds of medical technologies act to advantage or disadvantage specific social groups, Michel Foucault (1963) wrote at length about the medical gaze and how Western people's trust in medicine turns us into coercive subjects: uncritical and deindividualized. His term "docile bodies" refers to the manner through which subjects adopt the ideologies supported and enforced by the modern state. In the *Birth of the Clinic* (141) he writes:

> Over all these endeavours on the part of clinical thought to define its methods and scientific norms hovers the great myth of the pure Gaze that would be pure Language: a speaking eye. It would scan the entire hospital field, taking in and gathering together each of the singular events that occurred within it... . This speaking eye would be the servant of things and the master of truth.

There is little new in connecting panoptical observational regimes and Western biomedicine. But what has changed in relation to the principles of telecare is what Foucault (1977:212) called the ability of disciplinary mechanisms to cross the "technological threshold." The decentralizing, democratizing aspects of surveillance technologies through the welfare regime therefore highlights examples where one can consider caring surveillance in more critical and reflexive ways. So the idea behind enlarging the concept of surveillance is to reemphasize that there are few clear-cut divisions that distinguish power from care (Wood 2003). Self-surveillance thus becomes part of the obligatory care of the self. Such care accepts the responsibilities of constituting oneself as an acceptable and normativized citizen.

From this, it is worth noting that few in-depth qualitative studies of how users perceive telecare exist to date (Barlow, Bayer, and Curry 2005). Levy and associates (2003), for example, examined the responses and perceptions of older people to home-based technologies and to telecare in particular. The attitudes of a small group of participants who had no prior awareness or engagement with telehealth equipment were compared with the experiences of another small group who had used a video link in a remote consultation with their general practitioner. Participants took part in semistructured interviews in their own homes. Following the analysis of interview data they drew up and distributed a questionnaire to three client groups. These groups were as follows: day hospital patients, residents of a private housing association, and residents of a local authority-sheltered housing scheme. Analysis of the responses suggested that the main demographic factors associated with positive attitudes to telecare were age, home ownership, dwelling type (sheltered housing or not), and household composition. Their results suggest that the most likely mechanism by which to encourage older people to accept and use domestic systems are nurse-led telehealth services because such schemes facilitate trust between clients, providers, and the technology.

Participants for the present interdisciplinary project came from a group of several hundred cardiac patients in South Yorkshire. Research collaborators at the Barnsley District Hospital[1] recruited forty-five people who fell into a telecare intervention group and a nontelecare control group. A subset of thirty-two participants was chosen from them, mostly from the intervention group. The basis for this selection was the type of housing in which they lived—for example, Victorian terraced house, council house/flat, properly built or converted dwelling, detached or semidetached bungalow. Hence, at least five participants for these five different housing types were included, a number that is roughly representative of the average socioeconomic conditions of this area of South Yorkshire.

Of the thirty-two participants, there were twenty-eight men and four women. The gender bias is indicative of this medical condition that disproportionately afflicts men. The interviews were conducted in batches between early October 2005 and March 2006. These informal, semistructured interviews lasted about an hour and a quarter. A couple took over three hours. The interviews were conducted before the installation of the telecare package to get a preliminary sense of what the tenants expected of the project, their views of modern technology (particularly health technology), and their reasons for volunteering to participate. The interviews concluded by drawing up a sketch plan of the room in which the interview took place and an approximate plan of the entire home, more detailed if the participants offered to provide a tour. The idea here was to get them to speak about their domestic usages of space. A few people offered to sketch their homes out themselves which was ideal ethnographically since it tells us something about how they view their own living space.

Seventeen participants accepted the lifestyle monitoring sensors that meant that their homes were kitted out with various types of sensors for movement, occupancy (chair, bed), and utility (kettle, fridge, microwave, cupboards). Consequently, nearly half refused these sensors. Except for the five controls, all received a Docobot health monitoring computer with weight scale and blood pressure gauge (Figure 1).

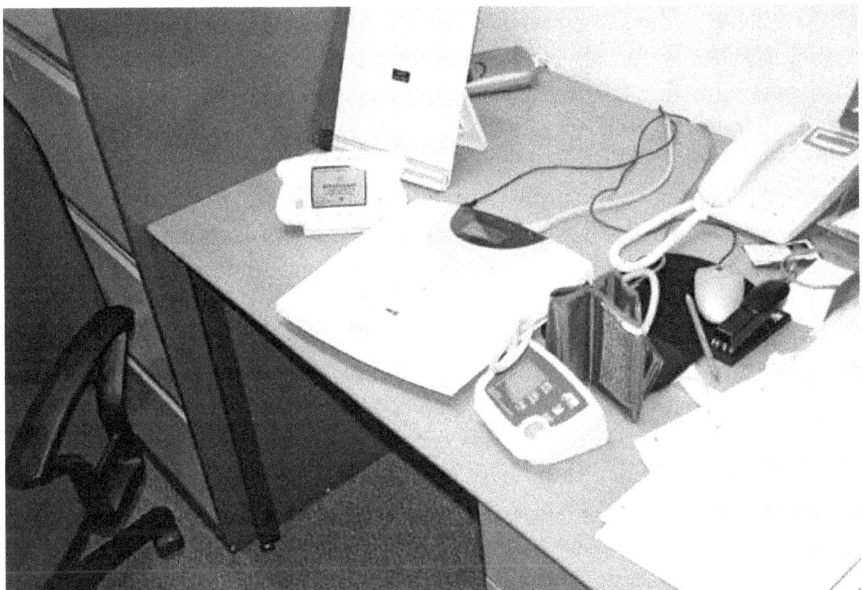

Figure 1: Telecare kit (left to right, Docobot, weight scale, blood pressure gauge).

The second batch of interviews with nineteen of the thirty-two participants were conducted nine months after installation from late November 2006 to January 2007. Because of the smaller number of people in this second group, the five controls were not interviewed. Also several deaths (six) occurred over the nine months and another individual withdrew because of health concerns. Finally, one couple declined a second interview because of complications surrounding their domestic relocation.

On average these follow-up interviews lasted for about two hours and involved questions about living with the system. After a series of questions about their homes, and their perceptions of the pros and cons of telecare, I asked the informants to take me through their daily routines and the uses they made of various technologies as well as their habitual places in their homes and beyond (Highmore 2002; Michael 2006). Not only was this to assist in the understanding of everyday lifestyles but also to correlate with whether the sensor data bear any resemblance to what people say they are doing.

AT: Answers and Tales

A perfect example of grievances of the type that sees hospital care as problematic comes from Mr. and Mrs. Lister,[2] who are in their early seventies. They are among the few participants to live in the heart of the city of Barnsley. They have lived in the same two bedroom house for the past 35 years. Prior to that, they say, "we were only up the road, in a house that got condemned so we had to move quick." They do not have any children but have extensively adapted their home with a complex network of ceiling-mounted railings for their three surrogate children, that is, their cats. Mr. Lister used to work as the manager of a construction company. He had his first and only heart attack 20 years ago in 1988. He went into hospital just in time after feeling ill all morning. He was kept in the hospital for 3 weeks and returned to work after a month. He accepts without doubt that the cardiac arrest center was excellent and saved his life. Indeed, his rationale for participating in this telecare project is founded upon a moral obligation to repay this debt toward the UK's National Health Services (NHS) in whatever way he can.

Yet in many ways this couple was exceptionally cynical about the modern health care system. They described how the doctors in Barnsley sent Mrs. Lister home to die in 1968 when she was deemed terminally ill (she does not suffer from CHF). After a few years of treatment and a referral to specialists in Sheffield, at the very time when the Barnsley Beckett Hospital had moved to its current location at the top of the hill overlooking the town, she remembers "telling the doctors at the time, I'm not coming back to Barnsley Hospital because you bury mistakes here." The couple went on to explain that the city used to have a lot of power cuts in the past because of the surrounding colliery activities. So living near the old Beckett Hospital in the center of town was convenient because the hospital would not be subjected to such outages. In Mr. Lister's words,

> But when they moved up Pogmoor, new hospital got a bad name early on. They seem to spend more time and money developing waiting areas. And referrals seem to be something they're good at. Must get paid by number of visits and referrals. They like you to come back...but the more and bigger hospitals get, the more people fall through the cracks and you end up with many incompetent specialists...these days you don't actually get cured.

The couple's complaints effectively relate to how patients get caught in a self-perpetuating system of nurturance where they are shuffled around, where one is never seriously ill but never altogether alright either. These perceptions would suggest that the Listers would be rather supportive of telecare since it sidelines the source of their grievance, the institutional hospital, from therapeutic experience. This only seems to be half the case, however. They have no problem with the daily use of blood pressure measurements, weight recordings or the Docobot computer for answering health questions. Nor did they feel that the series of interviews that they went through over the course of their participation was invasive. Quite to the contrary, the first interview with them lasted for over 3.5 hours. Nevertheless, they were considerably opposed to the installation of any sensors or lifestyle monitoring into their home. "I'm not too impressed with that," Mr. Lister mentioned. "It's invading on Irene's privacy too much and she doesn't like it."

It is this aspect of lifestyle monitoring through domestic sensors that most participants found difficult to accept. Indeed out of the thirty-two initial interviewees, only seventeen have had some form of LSM installed. In most cases this was not even a complete package either. Several have objected to having devices placed upon kitchen appliances and some have opposed fitting bed or chair sensors. Others still were reluctant about accepting the general PIR motion detectors, often because there is too much "exterior activity" in their home. But often implicitly there was a concern about the disturbances to décor or comfort as well as the invasion of privacy.

An interesting anecdote worth recounting is that Mr. Lister drove me back to my hospital accommodation after one of our interview sessions. On the way, he commented that when he has a hospital appointment, he will nearly always plan to eat in the Beckett café because the food is so good. Without being solicited by any of the questions in the interview, this high estimation of the canteen arose in discussions with several interviewees. What this raises more generally is a significant concern with regard to the mobility issues surrounding telecare assistance for older people. If going into hospital for a cardiac checkup with the anticipation of visiting the café becomes a meaningful break from the everyday existence of many participants, then an initiative for maintaining independence—which means that people remain at home more often—becomes slightly paradoxical, if not problematic. Such instances might further reflect the idea that even though most people complain that they do not like going to hospital even for routine checks, the setting can nonetheless provide an escape from familial settings or habitual everyday lives, at least for certain people.

On the whole, however, it was clear that many informants felt that the most useful thing to come out of any telecare participation is that it can save travel time for them. As Mr. Kinner (aged 77) says, "it's a good idea if it will reduce my visits to hospital because I seem to be a regular customer there at the moment." Most people also strongly supported the idea that if it helped make the NHS more efficient and saved money and time for health care professionals, it would surely be a good thing. But most importantly, even though they all had some level of self-interest for partaking, most nonetheless said that to some degree it was their ethical duty as concerned citizens to help the NHS research community by participating in this trial. The fact that many of them had been servicemen in the Second World War might help explain this moral obligation to the nation.

Another common source of concern for many of the informants relates to not being competent enough to use the technology involved adequately. Several suggested that the system needed to be "idiot proof." Yet this was frequently followed by the proviso that their generation did not grow up with the same proliferation of modern gadgets. Take the following exchange with the Mulders (in their late sixties) at the question: "So how do you describe your attitudes toward technology, toward electronic household devices for instance?"

> Mrs. Mulder: Confusing.
> Mr. Mulder: Yeah, we're getting up for 69, 70 and the developments and industries keep going faster and faster at it really. It's just a question of accepting them, that sort of thing. But if we'd done this 20 years ago, I would have been on top of it. It's like everybody says, our grandchildren come down and work out the video unit better than we can.
> Interviewer: But now you've managed to find out how those kind of things work, right?
> Mr. Mulder: Yeah sure. Once it's installed in the brain like, the brain cells aren't dying completely. I can remember stuff, haven't got to the stage where I've forgotten what happened the day before yet.

The Mulders have lived in the same semidetached three-bed house for 40 years. They are far from unique in thinking that the evolution of electronic technology has occurred at an unprecedented pace that has largely passed them by. This is effectively a reflection upon not having had the right exposure to deal with such things, that is, the generational construction of a technologically savvy habitus (Bourdieu 1977). One could thus say that the domestication process of such technology is slowed down. Yet it should nevertheless be pointed out that the issue of class is perhaps a variable that is being taken for granted by this prefacing of time. Even though they acquired their house privately (an exception to 75 percent of our informants) and put one of their two children through college, they were more at ease classifying themselves as "working class."

Perhaps more obviously fitting into this category will be Ronald and Jennie Pinter, who live in a nearby Barnsley suburb in a semidetached three-bed house with its large garden at the back. They bought the property (built in the late 1960s) from the council soon after "the right to buy" scheme was introduced under the 1980 Housing Act. As an aside it is worth pointing out that the groundwork for this scheme had been laid in the 1970s under the reign of the Labour government but Thatcher initiated the historic concept in 1980 with the famous statement: "We will help every Council Tenant to become Home Owners." Thatcher was of course also behind the closure of the coal mines, an industry that many participants such as Mr. Pinter have worked in.

Like most other informants, the Pinters' children have moved out some time ago but live locally and assist with household chores from time to time. Mr. Pinter is now 69 and had a triple heart bypass a few years ago. He is quite immobile and completely dependent on his wife, who complains that despite her own impairments related to aging, she now has more work than ever with the increasing demands of looking after her ailing husband. One of their main reasons for participating in the project is the desire to alleviate some of this burden of caring at this age. But the paradox is that Mr. Pinter does not feel completely confident that he will be able to use the system without further assistance, thus inflicting additional encumbrance upon his wife. As he says,

> There's only one concern I've got. I'm as thick as two planks normally. I'm not sure I'll be able to follow if it's too complicated. She'll [the district nurse] have to leave me a clear and simple list of what to do step by step, otherwise it'll be no use y'know... And my days [of] ill[ness] are increasing and if she [his wife] has to get involved with everything I start with in the morning, what I do in a morning till through the day, she'll never have time for anything else... So they'll have to leave me a really clear list of what to do and how.

Mr. Pinter uses the stairs only twice a day, to and from the bedroom. They have a toilet upstairs for the night and one on the ground floor for day use. He is no longer able to comfortably access the garden that he no longer frequents. They do, however, go out together to one of the local social clubs once or twice a week. Even though he says he now spends more than three quarters of his day in the same chair, he adds that despite his wife's objections, he will consider moving into a ground floor bungalow if his mobility gets worse. "We could get a stair escalator I guess, but only if the council or the NHS pay for it. If not, makes more sense moving, really." Here the tension between being a burden on his carer and his desire for a greater independence from her is even more wound up because she herself strongly opposes any relocation.

220 Laviolette

In this particular case, the discourse about burden ended there. But in many other instances, the notion of burden extended beyond the level of kin. Perhaps the place where citizenship was most overtly apparent was in dealing with the views expressed by certain people like Mr. Littlerock (aged 67) who suggested that ultimately he is worried about becoming a burden on the state. Mr. Littlerock has lived alone in his two-bed bungalow for several years since his wife died. Two of his three children live locally. "Hospitals would make you into an invalid if they could," he says while adding that he does not understand the reason since they are so strained for resources. He is very active and wants to die in a manner that is quick and dignified "like that Yorkshire gardener Percy Thrower. Y'know, second heart attack, bam! Gone." In other words, Mr. Littlerock does not want a protracted, drawn-out death. He goes on to say that he is not keen on survival for survival's sake, "even if the technological means are there to keep me ticking, that's not the point, y'know. I mean, there's enough old buggers like me around and the world doesn't really need more people who can't look after themselves properly." The discussion then naturally meandered to such topics as euthanasia as well as prolonging the longevity of life through pharmaceutical developments. He reads up to three newspapers a day and so is well informed about what the media have to say about these issues.

IT: Intimate Technologies

Some of telecare's awaiting solutions are indeed technological, but only in the widest sense of the word. That is, its incompleteness exists on more than technical grounds, although certain such barriers are still present. A significant development component in the area of research design for implementing telecare remains on methodological and conceptual levels of integrating comprehensive interdisciplinary techniques and frameworks. Pfaffenberger (1988:249) reminds us that the uses of technologies express social visions and "engage us in a form of life." It is their mythic and social facets that make technologies what they are, a form of enchantment perhaps (cf. Gell 1998). Hence, to understand the potential uses of new technologies, we need to draw out the connections of how they relate to various social visions, that is, to unpack their mythic and cultural dimensions.

In this sense, the implementation of telecare involves a more thorough, sensitive, and nuanced understanding of the architectural visions and technologies involved in designing homes that are to be adequately dwelt in. The home obviously becomes a more important and frequented space in the context of aging because other habitual spaces (predominantly the workplace) are removed from everyday interaction. There has been much debate recently about the notion of the "smart home" in relation to dealing with the ways in which the domestic sphere can be made friendlier to the needs of older people (Peace and Holland 2001). This is an interesting and socially significant discourse I would like to build upon in terms of including more illusive types of intelligences dealing with poetic, intuitive, and intimate experiences of space and technology.

One remarkable example in this respect is derived from my very last interview. This was with the Isaacses, a couple in their early seventies. I walked up to their house to be greeted by an unusual sight, a "For Sale" sign. This was slightly disconcerting because I had recently coauthored a paper entitled "Home Is Where the Heart Stopped" that argues that our participants are highly attached emotionally to their homes and they generally do not seem interested in moving despite having had life-threatening health issues there (Laviolette and Hanson 2007). However, here was an exception. Regardless, the couple in question was far from interested in speaking about telecare. Several times they were more engrossed with the activities of their budgie that was actively and noisily chirping away in a cage.

At the end of the interview they offered tea and biscuits and started talking about their newest hobby and infatuation, their building of miniature doll houses. Both husband and wife are equally involved in the process of fabricating, finding furnishings for, and eventually finding homes for their miniature abodes. These miniature houses are nearly three feet tall and take up considerable space so they never have more than two in their house at a time; they gift the others to children, grandchildren, and extended family. They are especially proud of the exceptional interior furbishing and fittings. Wallpaper, kettles, flower pots, rubbish bins, everything down to the finest details seem to be provided for. I resisted the temptation of asking if they planned on installing any form of telecare into these small virtual dwellings to make them even more realistic.

Miniaturization is one of the interesting conceptual themes to emerge here and of course the parallel that it has with attachment, emotion, and intimacy since this couple is, after all, attempting to sell their home because it has become too large. Instead, they hope to find a smaller home and a more intimate and manageable one. Previous research indicates the ways in which older people increasingly vacate their homes and inhabit smaller and smaller spaces (Metz 2000). In terms of its connection to the home and technology, the word intimacy originates from the Latin *intimus* that refers to the interior, the deepest thing within. Intimacy and being (intimacy of the self) are thus strongly linked, particularly in the context of "being within" or dwelling. This allows for the very process by which domestic contemplation and activities can literally and figuratively take place. That is, it provides the arena whereby the possibility for contemplation or practices can safely manifest themselves spatially. In Old German the word *buan* for "I am" also means "I dwell." Herein lay Heidegger's (1971 [1952]) fascination with Being through dwelling. Heidegger is interesting not only in terms of his writings on dwelling but also in the way he links this notion with two other issues, one at the heart of gerontology studies, time, or temporality; the other central to the discourses surrounding telecare—technology.

This personal communion of inhabiting the self, of in-dwelling, circumscribes an aura of intimacy. Here we can talk about the domestic appropriation of spaces of consciousness, contemplation, and activity. Such a conceptualization of the intimate interior invokes the relationship with the spaces that exist between the limits of both house and body, or as Pink (2004) might say, the "sensory home." Hence the existence of popular house/body analogies (i.e., intestines, again a derivative of *intus*, interior). Indeed, the phenomenological position applies particularly well when thinking of the home through the conceptualization of the body. In this sense, intuition—knowledge from within the body—is a concept that is inherently connected to dwelling, intimacy, interiority, and contemplation (Serfaty-Garzon 2003). Such a stance is in keeping with Warnier's (2006) recent musings on the nature of surfaces and containers.

What was fortuitous about asking people about their daily routines is that it is the daily, regular monitoring of one's own health that seems most beneficial to these outpatients. The routine health check, monitored both by specialists, remotely, and by our participants, making them more aware of their health, has yielded significant satisfaction in the rehabilitation process of these Barnsley residents. Telecare thus facilitates and provides an infrastructure for home-based daily checkups. It is therefore largely about making certain aspects of biomedicine mundane and habitual. Its strength lies in making medicine enter the intimate spaces of daily living.

Conclusion

One of the important social dimensions of wireless communication technologies is that some people feel that the immaterial auras of wave signals and bandwidths impact the very fabric of existence (Tacchi 2004). The creation of such a virtual/cerebral space is both problematic and equally seductive as an idea. Here the boundaries between freedom and restraint, mobility and invasiveness, security and threat, utility and dysfunction are blurred and weak. What becomes apparent, hence, is that many technologies like telecare are not only multifaceted but are frequently inherently ironic, contradictory, and paradoxical as Arnold (2003:234) has argued about the mobile phone: "What seems to be at work here, are phenomena that are not susceptible to a linear logic of cause and effect, situation and implication," not to mention how use of a mobile phone remakes the boundary between public and private spaces. The mobile phone is of course a relevant technological comparison because of its emancipatory potential for use in cases of emergencies as well as easier contact with friends and family. It is also part of a growing debate over the benefits and concerns for innocuous surveillance through Global Positioning System (Dwolatzky et al. 2006). Such ethical issues should hopefully begin to resonate with those for whom the long-term consequences of mainstreaming telecare has been an unreflexive process.

This raises a series of telling questions in the hypertechnological age (Silverstone and Hirsch 1992). How do we generate constructive applications and experiences for such ubiquitous networks and technologies? How do intimate technologies transform our selves and the way we construct narratives, relate to each other, play, dwell, work, and grow older? What is at stake in such questioning is to think about how to avoid the trap of telecare and LSM becoming virtual failures, or failures of the virtual as it were. By placing home-based technologies of well-being such as medication and telecare into the realm of "everyday intimacy," I hope to situate the debate as one concerned with familiarity, embodied experience, and an overall landscape of care.

Intimacy is a crucial element of domestic life and many interactive technologies designed for other purposes have been appropriated for use within intimate domestic spaces (Barlow et al. 2005). Yet a deficit persists in current understandings of how technologies are used within intimate relationships and how to design technologies to support intimate acts (Lyon 2001). This project attempts to address some of the issues that surround these deficits. I have used contextual interviews and other ethnographically informed techniques to examine how interactive technologies are used within both intimate as well as intimidating spaces and relationships (Andrews et al. 2006).

From this empirical case study, I would appeal for the provision of a comprehensive overview of the use of interactive assistive technologies to support the intimate act of domestic medical care. The purpose of such understanding would be to eventually inform the design of age- and condition-sensitive technologies that are suitable for appropriate homes. At present, however, it is fairly obvious that telecare is still a priori a clinician's technology. It has not yet been properly examined in the qualitative sociological context of how it relates to everyday domestic use and design. Rather, it gives epistemological privilege to the surface gaze, that is, where people go, whether they are eating, how much weight they put on, what their blood pressure is, and so forth. Yet we also need to account for the views encapsulated by one respondent, in his mid-eighties living alone, for the question: "You said when I walked in that you've been very satisfied with the trial, so what benefits do you feel that it's brought to you most?": "well...I feel as though I'm doctoring myself."

If domestic healthcare support, in the form of telecare, is to be medically beneficial to society in general and to its increasing number of older citizens with chronic conditions in particular, it must possess some of the empowering characteristics of what Mann (2004) calls "sousveillance" versus surveillance, or what others have termed the synoptical that stands in opposition to the panoptical (Haggerty and Ericson 2000; Hier 2003). Put differently, residential health monitoring technologies need to demonstrate a significant potential for vernacular, grassroots emancipation. This enabling potential would have to outstrip the more sinister controlling effects entailed by certain institutionally conceived forms of the medical gaze (Samuelsen and Steffen 2004; Sinha 2000).

In this sense alone, telecare would remain a formulaic monitoring or surveillance system for practitioners. But it is being appropriated by the emancipation of the routine health check by many of our CHF users. Such paradoxical facets are interesting and are only beginning to be adequately dealt with. We can therefore speculate that telecare demonstrates potential sousveillance characteristics even though the present level of supporting the independence of older people at home is still itself largely in its infancy.

This concept of sousveillance is useful insofar as it helps us nurture the significant vernacular power in the images and gazes that are encountered at the prosaic levels of the everyday (de Certeau, Giard, and Mayol 1998; Lyon 1994). Resistances form around such processes, which are in themselves part of this mechanism for conforming, normativizing, and standardizing behavior. So Foucault's adaptation of the model of the Panopticon suggests that it homogenizes power that then turns in on those who are aware of its presence.

We can thus speak of an "internalized ideology": a pure architectural and optical system of political technology that exists everywhere in everyday life. This chapter has sought to locate these institutionalized and vernacular sites of technologically enhanced power. It is part of readdressing what Lyon (2001) has recently argued is the unjust neglect on the part of social researchers to understand the issues surrounding caring surveillance. Of course in the case of home-based telecare assistive technology, the very site of the hospital is what is ultimately put into question, or more appropriately, what needs to be fundamentally reconceptualized. It is here that we can speak about developing residential landscapes of care. And in such spaces one can even allude figuratively to the onset of a domestic healthcare system that ultimately foresees and advocates, not the birth, but the "Death of the Clinic."

Acknowledgments

This chapter derives from an EPSRC/EQUAL project entitled "Supporting Independence: New Products, New Practices, New Communities" that was initiated by a research consortium comprising Barnsley District Hospital Trust, Dundee University, Imperial College London, Pocklington Trust, Tunstall Group, and University College London. I am especially grateful to UCL's Principal Investigator Julienne Hanson for her guidance and support. Thanks also to the organizers and participants of two conferences where earlier drafts of this chapter were presented: the Technology and Citizenship Conference at McGill University in June 2006 and the thirty-ninth international Social Policy Association conference entitled "The State of Welfare" at Birmingham University in July 2006. Thanks finally to Roland Littlewood for the invitation to contribute a full-length draft to the Medical Anthropology Research Seminar at UCL in February 2007.

Notes

1. H. Aldred, S. Brownsell, and M. Hawley (Research & Development Unit, Barnsley District General Hospital Trust).
2. Participant names are pseudonyms.

References

Andrews, Gavin, Robin Kearns, Pia Kontos, and Viv Wilson. 2006. "'Their Finest Hour': Older People, Oral Histories, and the Historical Geography of Social Life." *Social and Cultural Geography*, 7:153–177.

Arnold, Mark. 2003. "On the Phenomenology of Technology: The 'Janus-Faces' of Mobile Phones." *Information and Organization*, 13:231–313.

Barlow James, Steffen Bayer, and Richard Curry. 2005. "Flexible Homes, Flexible Care, Inflexible Organisations? The Role of Telecare in Supporting Independence." *Housing Studies*, 20:441–456.

Bourdieu, Pierre. 1977. *Outline of a Theory of Practice*. Cambridge: Cambridge University Press.

Certeau, Michel de, Luce Giard, and Pierre Mayol. 1998. *The Practice of Everyday Life*, vol. 2, *Living and Cooking*. Minneapolis: University of Minnesota Press.

Dant, Tim and V. Gully. 1994. *Coordinating Care at Home: A Handbook for Organising Support for Elderly People at Home*. London: HarperCollins.

Department of Health. 2000. "National Service Framework for Coronary Heart Disease." Available at: http//:www.dh.gov.uk.

Dwolatzky, B., E. Trengove, H. Struthers, J. McIntyre, and N. Martinson. 2006. "Linking the

Global Positioning System (GPS) to a Personal Digital Assistant (PDA) to Support Tuberculosis Control in South Africa: A Pilot Study." *International Journal of Health Geographics*, 5:34.
Feyerabend, Paul. 1978. *Science in a Free Society.* London: New Left Books.
Fisk, Malcom. 2003. *Social Alarms to Telecare: Older People's Services in Transition.* Bristol: Policy Press.
Foucault, Michel. 1963. *Naissance de la Clinique: l'Archéologie du Regard Médicale.* Paris: PUF.
——. 1977. *Discipline and Punish: The Birth of the Prison.* London: Penguin Books.
Gell, Alfred. 1998. *Art and Agency: An Anthropological Theory.* Oxford: Oxford University Press.
Haggerty, Kevin and Richard Ericson. 2000. "The Surveillant Assemblage." *British Journal of Sociology,* 51:605-622.
Heidegger, Martin. 1971 [1952]. *Poetry, Language, Thought.* New York: Harper & Row.
Hier, Sean. 2003. "Probing the Surveillance Assemblage: On the Dialectics of Surveillance Practices as Processes of Social Control." *Surveillance and Society*, 1:399-411.
Highmore, Ben. 2002. *Everyday Life and Cultural Theory: An Introduction.* London: Routledge.
Kuhn, Thomas. 1962. *The Structure of Scientific Revolutions.* Chicago: University of Chicago Press.
Latour, Bruno. 1987. *Science in Action: How to Follow Scientists and Engineers through Society.* Milton Keynes: Open University Press.
Laviolette, Patrick and Julienne Hanson. 2007. "Home Is Where the Heart Stopped: Panopticism, Chronic Disease and the Domestication of Assistive Technology." *Home Cultures*, 4:1-20.
Law, John and Annemarie Mol. 2001. "Situating Technoscience: An Inquiry into Spatialities." *Environment and Planning D: Society and Space*, 19:609-621.
Leach, Edmund. 1968. *A Runaway World: The 1967 Reith Lectures.* Oxford: Oxford University Press.
Levy, S., N. Jack, D. Bradley, M. Morison, and M. Swanston. 2003. "Perspectives on Telecare: The Client View." *Journal of Telemedicine and Telecare*, 9:156-160.
Livingstone, David. 2003. *Putting Science in Its Place: Geographies of Scientific Knowledge.* Chicago: University of Chicago Press.
Lyon, David. 1994. *The Electronic Eye: The Rise of Surveillance Society.* Cambridge: Polity Press.
——. 2001. *Surveillance Society: Monitoring Everyday Life.* Buckingham: Open University Press.
Mann, Steve. 2004. "'Sousveillance': Inverse Surveillance in Multimedia Imaging." *ACM Multimedia*, 3:620-627.
Metz, David. 2000. "Editorial: Innovation to Prevent Dependency in Old Age." *British Medical Journal*, 320:460-461.
Michael, Mike. 2006. *Technoscience and Everyday Life: The Complex Simplicities of the Mundane.* Maidenhead: Open University Press.
Peace, Sheila and Caroline Holland. 2001. *Inclusive Housing in an Ageing Society.* New York: Polity.
Pfaffenberger, Brian. 1988. "Fetishised Objects and Humanised Nature: Towards an Anthropology of Technology." *Man*, 23:236-252.
Pink, Sarah. 2004. *Home Truths: Gender, Domestic Objects and Everyday Life.* Oxford: Berg.

Porteus, Jeremy and Simon Brownsell. 2000. *Using Telecare: Exploring Technologies for Independent Living for Older People.* Bradford: Anchor Housing Trust.

Samuelsen, Helle and Vibeke Steffen. 2004. "The Relevance of Foucault and Bourdieu for Medical Anthropology: Exploring New Sites." *Anthropology and Medicine*, 11:3-10.

Serfaty-Garzon, P. 2003. "Le Chez-Soi: Habitat et Intimate." Pp. 65-69 in *Dictionnaire Critique de l'Habitation et du Logement*, edited by M. Segaud and J.C. Driant. Paris: Editions Armands Colin.

Silverston, Roger and Eric Hirsch (Eds.). 1992. *Consuming Technologies: Media and Information in Domestic Spaces.* London: Routledge.

Sinha, Arushi. 2000. "An Overview of Telemedicine: The Virtual Gaze of Health Care in the Next Century." *Medical Anthropology Quarterly*, 14:291-309.

Tacchi, Jo. 2004. "Researching Creative Applications of New Information and Communication Technologies." *International Journal of Cultural Studies*, 7:91-103.

Warnier, Jean-Pierre. 2006. "Inside and Outside: Surfaces and Containers." Pp. 186-196 in *Handbook of Material Culture*, edited by Chris Tilley et al. London: Sage.

Wittgenstein, Ludwig. 1969. *On Certainty.* Oxford: Blackwell.

Wood, David. 2003. "Editorial: Foucault and Panopticism Revisited." *Surveillance and Society*, 1:234-239.

15
The Zapper and the Zapped: Microwave Ovens and the People Who Use Them

Tina Peterson

Most people do not think that they ever give a thought to microwave ovens. That is to say, they do not have any specific thoughts of which they are consciously aware. Like much of the technology other authors examine in this volume, microwave ovens have become so mundane a part of our everyday material culture that they seem unremarkable. Indeed, most of my informants at first knitted their eyebrows in bemusement at my questions about what seems to be such a banal object in their lives. But as it turns out, once they felt comfortable and justified in reflecting on it, stories began to flow, techniques were described, theories were shared, and opinions were made clear.

The very fact that people have stories to share about mundane objects like microwave ovens and cautionary tales to pass on regarding their use does demonstrate that their presence in our lives is anything but ordinary. Many of us may remember a time before the appliances were ubiquitous in North American households; younger people may not recall a kitchen being without one. The mundane fact of life that is a microwave oven is noteworthy if for no other reason than the fact that it has become an inextricable part of our lives in a relatively short time.

In this analysis, I aim to make sense of the ways people use, think about, and interact with microwave ovens in their daily lives. The appliance has shaped, and been shaped by, the historical moment at the turn of the twenty-first century. Indeed, first, it was and is a star player in the industrialization of housework (Cowan 1983) and a kinder, gentler domestic application of technology initially intended for military purposes. Second, its development was an iterative process of negotiation between men and women working in traditionally gendered spheres; engineers and manufacturers (primarily men) depended on feedback from home economists and product testers (mostly women) to fine-tune the appliance's functionality (Cockburn and Ormrod 1993). Building on this and more, my primary research question is, what does the microwave oven's presence (or absence) in our everyday existence say about the lives we are leading? My focus, in particular, is on technicians (microwave users) and their techniques (i.e., how people use microwave ovens and why). I draw upon various perspectives and concepts outlined in this book, in particular symbolic interactionism (see Vannini chapter five) to understand the ways in which we materially interact with the appliance. First, it is useful to consider how and when microwave ovens arrived in our kitchens.

Historical Context

The story of how the microwave oven was invented has all the elements of a good "gee whiz" tale. According to Hammack (2005) an engineer named Percy Spencer was working with vacuum tubes in 1946 for the Raytheon Corporation. It is not clear exactly how Spencer made the connection between the tubes and their potential for heating foods, but the oft-repeated story is this: he had been standing near a vacuum tube in the lab for some time when he noticed that a chocolate bar in his pocket had melted. Later, he and his colleagues popped kernels of corn and blew up eggs by placing them near the tubes. Spencer experimented with the technology until he had a working prototype, a metal box into which microwaves (electromagnetic waves of a certain length) were directed. Food placed in the box became hot very rapidly as the water, fat, and sugar molecules in it were vibrated by the waves.

The earliest microwave ovens were designed for commercial use. They were enormous, heavy, and cost more than US$2,000 (the equivalent of US$20,000 today) (Hammack 2005). Various models for home use were introduced in the 1950s, but none proved popular initially. By 1970 manufacturers had hit their stride with the design and marketing of the microwave oven, and 40,000 of the appliances were being sold annually by Raytheon and competitors at a price of US$300 to US$400 each (ibid).

One of my informants described a demonstration performed in his seventh grade science class in 1962 by representatives from the Amana Corporation. They brought with them a microwave oven prototype that was about the size of a large modern countertop model and showed the students how it worked. Rodney described how fascinated he and his classmates were by it:

> It was cool because they could put a cup of water in there and it would boil and everything. This wasn't even on the market yet, and they didn't know what they were going to call it, but they *said* it was going to be called a "radar range." It had a light inside it, so you could see that the water was boiling, and then they'd take it out and everyone said, "wow, that is so cool!"

The appliance was adopted relatively quickly by U.S. consumers in the 1970s and 1980s. The statistics vary widely as to exactly how many homes owned microwaves and when they acquired them; an accurate count of how many were in use in households may be inferred by sales figures for the appliances. According to data on cumulative shipments of appliances, the speed of adoption of microwaves was comparable to the speed of adoption of electric refrigerators 50 years earlier. In 1973, penetration was 1 percent—in other words, 1 percent of households had the appliance. Within 9 years, penetration was 20 percent. Four years after that, 50 percent of U.S. households had one (Bowden and Offer 1994). In 2001, 92 percent of U.S. households had microwave ovens, a higher number than with air conditioning (80 percent) (U.S. Census Bureau 2007).

Many of the material properties of microwave ovens (touted by manufacturers of early models) continue to motivate consumers to buy them today, housing developers to include "built-in" models in new kitchens, and universities to provide them in dormitory halls. Conventional wisdom about the appliances is that, compared to conventional ovens, they are

- energy-efficient, as they heat and cook foods faster;
- safer for children to use without adult supervision;
- compact and more portable;
- easier to use, especially as specific functions such as "popcorn" are automated.

A common assumption is that even people who cannot actually "cook" can at least use a microwave, and thus enjoy a degree of self-sufficiency.

The appliance is now considered such a standard tool in U.S. households that it is included in children's play kitchens. Designed on a small scale as a model of a real life kitchen, a play kitchen is ostensibly meant to mimic a typical household environment for children's make-believe activities. In most play kitchens sold by a national toy store chain in the United States, the microwave's place is central; it is positioned either next to or above the stove (Toys "R" Us 2008). It is worth noting that many play kitchen microwaves have an image of a popcorn bowl on them, which presumably is meant to represent a child's understanding of typical use of the appliance. Along with dishwashers, refrigerators, ovens and stoves, the microwave has become de rigueur in American kitchens.

Theoretical Framework

I became interested in symbolic interactionism as a theoretical base for my ethnography because of comments many of my informants made early on in my research. As interactionists explain, relationships and interpersonal communication were key factors in the ways individuals made sense of microwaves and the account they gave about their reasons for using them. For example,

My sister sent me an e-mail about microwaves.

People say you shouldn't do that with a microwave, but I do it anyway.

I have this friend who knows someone who put her dog in one.

Because people's usage of and feelings about their microwave ovens are socially situated, it is useful and appropriate to use symbolic interactionism to guide my analysis.

Many of my informants' usage of microwaves is highly influenced by what they hear about the appliances from others; such behavior follows the classic model of symbolic interactionism described by Blumer (1972). However, relying only on this basic definition limits my analysis considerably. I intend to base my examination of my informants' stories about their microwaves on the concept of scripts for their performance that Vannini describes in chapter five (see also Kien this volume). Material objects such as appliances do not have immutable essences; as Vannini and Kien explain, their agency is shaped by the agency of the people who use them, and by the scripts they are endowed with and enact. I also intend to analyze people's material interactions with their microwave ovens according to the ecological perspective Vannini describes, one that includes temporal and spatial aspects, as well as self-indication and negotiation.

Method

Most ethnographers face the challenge of trying to observe naturalistic behavior among members of a community while at the same time being an unnatural and often intrusive presence in that community. It is even more difficult to observe behavior that people may be self-conscious about even when they are not being watched. Such is the case with microwave oven usage. Some of my subjects expressed discomfort regarding their regular use of the appliance, while others seemed to want to show off what they could do with it. Natural behavior often becomes self-consciously performative—arguably the opposite of the dramaturgical unawareness that some naturalist social scientists seek. For example, my own mother noticeably altered her microwaving habits after speaking with me about how she uses the appliance.

To minimize the disruptive significance of my presence as a researcher in the settings where I observed people and conducted ethnographic interviews, I tried to ask questions that were as open-ended—and thus resembling mundane conversation flow—as possible. Unfortunately, it is often difficult to elicit a response from someone about technology they consider unremarkable. When

I said, "Tell me about microwaves," the most common responses were shrugs, confused expressions and the question, "What do you mean? What do you want to know?" Most of my conversations with people developed very slowly, and many informants did not really begin to share stories until after the interview concluded and my digital recorder was tucked away. Some people seemed reluctant to talk in much detail until I shared with them some of the thoughts others had expressed. It was as if they were waiting to hear that it was socially acceptable to have, and say aloud, an opinion about such a banal thing.

I was led to people in my snowball sample via conversations with friends, acquaintances, and strangers. In the dozens of conversations, I encountered perhaps one or two people who did not have a story they wanted to share. I conducted semistructured interviews with two dozen people, and had conversations with many more dozens whom I encountered over the course of my research. The first person who shared her opinion about microwaves with me was a woman who, like me, was shopping for digital recorders at an office supply store. I mentioned the topic of my interviews, and she immediately began to tell me a story; she confided in me that she had long suspected that radiation from her microwave was the cause of many of her health problems. Also, a classmate shared stories about a microwave bread recipe gone awry, and a colleague told me that his aunt cooks everything in the microwave, a habit he considers "weird."

I have also observed others' microwave use casually for many years; every company I have ever worked for had a microwave in its employee lounge. For the purposes of this study, I observed people using microwaves in grocery store café areas and convenience stores. The former was a more amenable venue because, as I discovered, it is difficult to remain inconspicuous while standing still in a convenience store.

User Types

From conversations with people about their habits and beliefs regarding their microwave ovens, three basic types of usage begin to emerge. Savvy users are the most likely to understand (or believe they understand) how a microwave functions, and to think that it does certain tasks quite well and is useless for others. They are also the most likely to stand in front of the appliance and watch what happens to food—or other substances—inside it. A subset of savvy users are those who perform science experiments with the appliance; they are likely to put materials in the microwave that manufacturers' manuals explicitly direct users to avoid.

The second type of user is the reluctant habitual. Many such people speak about their microwaves with some degree of fear, yet they say they use the appliance at least once or twice daily. A common misgiving is that radioactive energy is permeating the food and their bodies if they stand too close to it. There is also the suspicion that the microwaves themselves alter the food in a negative way, removing nutrients or damaging proteins. Many reluctant habitual users speak in ritualistic terms about what containers they will and will not use in their microwave ovens: plastic is a no-no, as many believe that the microwaves alter the plastic components and release chemicals into the food as it heats. Many describe how they avoid microwaving paper or Styrofoam cups (reheating drinks is a common use), and instead prefer pouring their coffee or tea into a separate container specifically designed for use in the appliance.

The third type of user is actually a nonuser. This type includes (1) those who have used a microwave in the past and now reject it, (2) those who distrust it and want nothing to do with it, and (3) those who reject it on principle. Many avoiders were once reluctant habituals, but they now keep their microwaves in storage or in their garage and use them less than once a month. Other avoiders are reluctant habituals except when influenced by someone else in their household (e.g., a spouse or child) not to use the appliance to prepare food for that person. The most adamant of avoiders are foodservice professionals and "foodies" who believe using a microwave prioritizes speed of preparation over the quality of the final product.

Savvy Users and Mad Scientists

For savvy users a microwave oven is thoroughly domesticated. For mad scientists the level of domestication is so advanced that they will often rescript the performance of microwave ovens in search of a thrill and some suspense, or perhaps a demonstration of savoir faire. Savvy users generally express the least amount of uncertainty regarding microwaves; indeed, they may roll their eyes when they hear about others' fears of the appliance. While others might try to avoid being near the microwave while it is on, Savvy users often peer through the door to observe the action of the waves inside. Rodney described watching a bowl of instant oatmeal rise and fall in spots as the microwaves penetrated different parts of it. He also insisted that, while a microwave fails at many things—"I'd rather eat a cold slice of pizza than a piece of pizza that has been put into the microwave"—it is excellent for cooking bacon if you know how to do it. He went on to describe an elaborate technique involving many layers of paper towels that produces evenly crisp bacon slices.

Many Savvy users have a repertoire of dishes they make exclusively in the microwave. Geeta, a 29-year-old married woman who grew up in eastern India, said she uses her microwave nearly every day to cook rice and make fish or shrimp dishes. She said her parents bought their first microwave when she was 18 years old, and that many of her Indian relatives and friends have specific recipes for the microwave that they use regularly. In addition to reheating food in it, her mother uses it to make tandoori chicken, and Geeta said it turns out just as well as when it is baked in an oven.

Geeta owns an expensive electric rice cooker, but she uses it only when she has company over for dinner. When she and her husband are the only ones eating, she cooks a smaller amount of rice in a special plastic cooker in the microwave. Like other savvy users, she has mastered the idiosyncrasies of the appliance through trial and error: "In the rice cooker the rice will always turn out the same...but in the microwave, if you put in more water than is needed, the rice will be really soggy. So you have to be careful with the water." She describes her techniques in exacting detail, from the Pyrex containers that work best to the small pieces of fish that soak up marinade and cook evenly.

A subset of Savvy users are microwave mad scientists, people who do not think much of the appliance as a kitchen tool but love to use it as a laboratory. Evelyn, a 17-year-old who plans to major in chemistry in college, sees the microwave as rather useless for cooking but an endless source of entertainment. When asked what she uses it for, she responded,

> Making hot chocolate. That's my number one use for microwaves. And cup of noodles...and...trying to burn things. Putting tin foil in the microwave; it just never gets old. And, I mean, your food tastes bad for the next couple of days, but it's still worth it.

She went on to describe microwave experiments she observed in college dorms during a summer program. In "Peep wars," she said, two marshmallow Peeps were placed inside a microwave, stuck with wooden toothpicks poking out toward each other. As the carousel turned and the microwaves penetrated the Peeps, they expanded and pushed the toothpicks outward: "They start growing, and so then they're, like, jousting. Except they joust really slowly. They get bigger and bigger, and eventually one pops the other one. Jousting and then they just explode."

Evelyn is not alone in her fascination with microwaves and their effects on various objects not intended as food. A research engineer at the University of Washington maintains a Web site describing "unwise microwave oven experiments." The main page of the site contains a strongly worded statement warning children not to try at home the projects he describes. Some of the microwave demos he lists are "Lightning Storm," "Pyrex Magma," and

"Candle Spews 'Ball Lightning'" (Beaty 2007). One of my informants described with excitement how to put a CD in the microwave and watch the "lightning" fly around and make marks on the surface of the disc. According to Beaty's Web site, this is one of the classic "don't try this at home" experiments. Whether they use the appliance for gourmet cooking or blowing things up, Savvy users are comfortable with the technology and are aware of its functionality and limitations. They see a microwave as just another tool and are more likely to trade recipes and ideas for experiments than cautionary tips, a popular currency among the next group I will discuss.

Reluctant Habitual Usage and Urban Legend

Reluctant habitual users tend to be highly influenced by others' opinions and various health panics floating in popular discourse. They use a microwave oven regularly in spite of their own trepidation about the appliance. Such people tend to speak in normative terms when they describe their usage of microwaves; their comments include "I know I shouldn't do that" and "They say it isn't a good idea to..." Reluctant habituals also tend to use specific verbs when describing the functionality of the appliance. Microwave ovens "nuke" or "zap" foods, and the effects they have on food are far different than the effects of a stove, an oven, or any other more traditional kitchen appliance. Reluctant habituals' uncertainty and low comfort level with the technology may be related to the fact that they tend not to understand how a microwave oven works.

Hannah, a 32-year-old woman who lives alone and likes to cook, described her understanding of how microwaves function:

> I know I'm not supposed to put metal in it. Um...I know it's probably radioactive material and sometimes when I stand in front of it I think to myself, "I shouldn't be standing in front of a microwave." Um...I never feel quite comfortable putting food in it. Because I feel like it's being nuked. So I'm not psyched on the microwave, but it's very functional for the lifestyle that I lead.

Many other reluctant habitual users echo Hannah's sentiment that life moves so fast that it is unrealistic not to use a microwave for food preparation. While acknowledging that the appliance enables them to eat more efficiently and get on with other things, they also make a distinction between the kind of food preparation they do with or without a microwave oven. Many use the term "real cooking" to distinguish how they prepare food on a stovetop or in an oven in contrast to what they do with a microwave.

Merle, a 50-year-old mother of two, described how her cooking habits changed once she began to use a microwave:

> When they first came out, I thought they were the greatest, and as time went along, I realized I started, um, taking shortcuts with meals and things like that. And just stick 'em in the microwave, and, you know, everything would be done in a few minutes or whatever.

She went on to say that once she began using a microwave, she started buying canned or frozen vegetables instead of fresh ones, and her husband noticed the difference in her cooking. When it was her turn to cook dinner, she tended to reach for a takeout menu or use the microwave. Her husband "didn't particularly care for that, 'cause he...really enjoyed cooking, and he did everything fresh."

Many of my informants—of all types of users—made a clear distinction between using the stove or the oven and using the microwave to reheat or zap food that was already cooked. Like Merle, many made an association between microwaving and consuming convenience food or ordering takeout. Michael, a 28-year-old man who lives alone, described his infrequent use of the appliance:

> Depending on what the leftovers are, I use it to reheat leftovers, but usually that's, like, takeout food. I also use it for, like, bad convenience store food...that I get on occasion for, like, postdrinking days (laughs). But I don't use it for, um, "real cooking."

Reluctant habituals tend to pay close attention to what others tell them about the appliance—thus displaying a close attention to the dominant scripts regulating the performance and role of the microwave. Several people described e-mails they received from friends or family members persuading them to be cautious about their microwave use. One such e-mail—mentioned by three of my subjects—was an account of an experiment: two identical plants were watered (one with ordinary tap water and the other with the tap water that had been microwaved and cooled) and observed over time. The writer of the e-mail claimed that the plant that was given microwaved water had stunted growth, and he had included a photo of a side-by-side comparison of the plants. Merle is one of the participants who mentioned this e-mail, and she said she stopped using her microwave as much after she read that, and began heating food and water on the stovetop more often.

The mail about the "plant stunted by microwaved water" was just one of the many forwarded e-mails informants described as influencing their behavior. Tamara, a 29-year-old married woman, was noticeably embarrassed as she described how an e-mail from her sister finally convinced her to stop heating food in plastic containers in her microwave. Her sister often forwarded e-mails that made her roll her eyes, she said. The latest one described a friend-of-a-friend whose son had leukaemia, and his parents and doctor suspected it was a result of eating food microwaved in plastic. Tamara described her reaction:

> I was trying to blow it off. But then a part of me—that part of me that was little, that had the little worry—got really big. And I don't know—I snapped, and I was like, well, what if that's true? And then, it could be affecting us! And I have all this old Tupperware, I've had it for years, and I've microwaved it, and, you know, dishwashed it, and like, heated it up and cooled it down, and maybe it changed the form of the plastic and it's toxic.

She soon threw out all her Tupperware containers, and replaced them with glass dishes that she feels more comfortable using in the microwave.

Avoiders and Their Influence

People who make it a point not to use microwave ovens tend to be either influenced by others' opinions and various health panics (to the point that they radically change their behavior) or highly opinionated themselves regarding the appliance. Many avoiders have used a microwave oven regularly in the past but now live without one. David, a 34-year-old married man, said he and his wife did not bother taking their microwave with them when they moved to a new city. The appliance now sits in storage in his in-laws' basement. He said he and his wife do not miss having it in their kitchen, especially since it was only marginally functional. It often turned on unexpectedly when he closed the door, he said, and only occasionally did he manage to make it work correctly by pressing a certain combination of buttons. He also expressed a certain amount of disgust and fear regarding the appliance:

> They, like, get a stale smell in them from the food, and older ones—even if you wash them—the cooking process, like, makes that come out. It's in the plastic after a while, so...[sucks in breath] it's kind of gross. And I don't like opening the microwave. I'm glad I don't have one. I don't like them. They make me feel like I need to be wearing lead underwear.

Several other avoiders described keeping their microwaves in a garage or other space separate from the kitchen, and said they plug them in and use them once in a great while to make popcorn or defrost frozen meat. Most avoiders have a basic understanding of how a microwave functions, but they remain skeptical about the type of energy it uses and how it affects food.

Helen, 35, is the executive chef of a highly rated restaurant, and she has no microwave in her kitchen at work or at home. Giving the microwave an agency—and a devilish one at that—of its own, she thinks the appliance is "the devil's tool," and said the only thing it does well is make popcorn. Not surprisingly, her primary concern in food preparation is flavor, and she said a microwave simply cannot duplicate the caramelization process that happens when food is cooked in a metal pot using a traditional heat source. She says she does not know anyone who "really cooks" using a microwave, echoing the distinction several reluctant habituals made between microwaving and cooking.

If someone is in such a hurry that they cannot take a few extra minutes to prepare something on the stove, Helen says, they are clearly concerned more about time than flavor. She believes that people who do not take the time to cook properly are losing out on the kind of pleasure the Slow Food Movement promotes: a meditative, absorbing activity that rewards the individual with more than just a tasty meal. Helen firmly believes that people would reconnect with the pleasures of cooking if they let go of their obsession with quick preparation that microwaves afford: "I would wager that if people threw their microwave away, and actually started working without it, they wouldn't miss it." She also expresses concern about adverse health concerns of eating microwaved food, an issue she thinks should be considered more seriously. If someone invited her to dinner and served her food that was prepared in a microwave, she says she would be polite but a bit unnerved about the energy used to cook the food. She jokes, "I would choke down whatever they made me, and probably do a breast exam later."

Avoiders such as David and Helen may influence others to become avoiders or at least reluctant habituals. Many of the e-mail forwards and Web sites cautioning against overreliance on microwaves are likely authored by particularly vocal avoiders. The fact that most people do not understand exactly how a microwave works may make them more receptive to such cautionary tales. There exists an undercurrent of what may be described as neo-Luddite sentiment in such narratives. One reason people may distrust such technology is the fact that they cannot directly see how it works. A burner on a stove becomes hot when it is turned on, wood and charcoal burn, and we can observe through our five senses what happens to food when it is exposed to such heat sources. What goes on inside a microwave is less obvious. We hear it hum, we see the light go on, and we can hear the water boiling or the corn kernels popping. Yet we cannot directly observe the point of contact between the food and the energy source that causes such reactions. More than one informant explained their distrust of the appliance: "It just doesn't seem natural."

Scripting Microwave Use

Driven by my interest in the ways people cook, I have casually observed some who have been using microwaves for years in office settings and in the homes of my friends and family. Unfortunately, as soon as I mentioned my project to people I knew, many of them became overtly aware of their microwave usage in my presence and began to radically alter their behavior. The reluctant habitual users became even more reluctant, the savvy users' enthusiasm grew as they showed off all that their appliances could do, and the avoiders became downright preachy in their admonitions not to use the device. It was as if they had established roles and scripts (see chapter five) for the appliances as well as for themselves, and became aware of the need to follow them even more closely when observed by others.

Most of my formal observations were done in the café area of an upscale grocery store in an urban neighborhood. I spent a total of 15 hours over several weeks at different times of day observing people. People who use the space generally eat and drink things they have purchased in the store. Near the eating area there is a counter where a small microwave sits next to a toaster, bottled condiments, and baskets of disposable cutlery.

Most of the microwave users I observed were those who eat alone. Many of them heated up beverages in paper cups, and others heated up soup or deli items they had purchased. Some stood in front of the appliance as it hummed, and fidgeted with things they had with them while they waited for their food to heat up. Several opened the door of the appliance numerous times without turning it off, and stirred or tested the temperature of the food or drink before putting it back in and restarting it. Only a handful of people who used the appliance sat down to eat with others; many had their grocery bags with them and stopped to use it to heat their drink before they left the store.

Nearly all the microwaves I have seen in convenience stores, grocery stores' café areas, and other public spaces had some sort of stain or food residue inside them. Several of my informants mentioned disliking having to use public microwaves to heat food because they tend not to be clean (interestingly, those same informants admitted to not cleaning their own microwaves very often). It is possible that their discomfort is a symptom of the modern condition in which technics such as affordable cars render common spaces such as public transit unnecessary and avoidable for many. As our technology becomes more personal (cell phones versus a landline phone, for example), we tend to become wearier of others' physical presence and material objects they leave behind. It is worth noting that near the microwave I observed in the store there sits a dispenser of disposable sanitizing wipes.

When microwave ovens were first sold as home appliances, many manufacturers included a cookbook in the box. These cookbooks featured recipes that, in hindsight, most long-time microwave owners would find ridiculous. Samsung's 105-page Microwave cookbook, which was included with models sold in the late 1970s, welcomed new owners of the appliance to try recipes such as veal cutlets cordon bleu, chilli cheese corn bread, and fresh peach pie. At the back of the book is an appendix of instructions for cooking common frozen and convenience foods such as pot pies and "pouch dinners."

Experience quickly taught many users of this new technic what the microwave is and is not good for. Bread recipes rarely work, as the outer edge of dough becomes tough and leathery while the inside remains raw. Most meat and poultry needs to be browned in a pan before being put in the microwave, and then rotated and flipped just as often as when it's cooked on the stove or in the oven. For many of my informants, the microwave began as a novelty in their kitchens but quickly became a one-or-two-trick pony, useful primarily for warming leftovers or making popcorn.

Accessories intended to enhance the appliance's functionality first appeared in the late 1970s and early 1980s, and continue to be sold by some kitchenware retailers. Plastic egg boilers, rice steamers, bacon racks, and potato stands are all designed to maximize the contact of the microwaves with the food. Yet most of my informants—both savvy and reluctant habitual users—maintain that the appliance does only a handful of things well.

Manufacturers of the appliance have clearly paid attention to their customers' habits, because the vast majority of microwaves now available feature special buttons for pizza and popcorn. These buttons, as it were, script users' behavior, but do it on the basis of what lessons manufacturers have gleaned from users themselves. In addition, many new models of microwaves now also feature additional functions. Some newer "hybrid" models broil and toast food as well; such functions are presumably in response to complaints from users that meat needed to be prebrowned and breads became tough and chewy when heated with microwaves. The scripts that accompany this, or arguably any other technology, are thus the obvious productions of an ever-expanding ecology of relationships and the changing definitions of a technic's role and repertoire of performances.

Discussion and Conclusion

Following on the idea of increasingly personalized dimensions of technoculture, I was struck by the solitude my informants described as the context of their microwave use. Very few described eating with other people after microwaving a meal or a drink. In the one or two stories in which a microwave was narrated to have been used to prepare more than one meal, it was seen as insufficient for the task. One woman told me about a social gathering she had with some friends, where large bowls of chilli were to be served buffet style. The owner of the gallery where the event was held provided two microwaves for the food to be reheated on-site. The appliances failed spectacularly at the job, causing intermittent electrical failures in the gallery and leaving the food mostly cold.

My observations that most people ate alone after they used a public microwave reinforced a theme that ran through many of my informants' narratives. Many said that when they heated leftovers, it was on plates one at a time. One woman said she often leaves a plate of food for her teenage son in the microwave, knowing he will eat dinner whenever he comes home. This tendency toward individual mealtimes and eating alone was foreshadowed in a 1974 magazine advertisement for a Litton microwave oven. A woman and two "clones" of herself are shown carrying platters of three different meals and heading in different directions. The headline of the ad reads, "for the woman whose family eats in shifts." The fact that microwave ovens work best at heating or cooking one item at a time means that people who use the appliance before sharing a meal with others often cannot eat at exactly the same time as their companions. In my casual observations in offices where I have worked, friends who wanted to eat together would often have to take turns using the microwave to heat up their food, and then one would eat while the other waited for their food to finish. It is noteworthy that among all the accessories designed to maximize the functionality of a microwave oven, I never encountered a product that enables two plates or frozen entrées to be heated at the same time. Solitude and self-sufficiency are two characteristics of the performance of a microwave that have clearly emerged with users' choices (cf. Lemonnier [2002]) over time.

In terms of its ecological significance in temporal and spatial terms, a microwave oven is the opposite of a fire pit or hearth. An open fire, an oven, or any other mechanism by which something is roasted is commonly a communal space around which people gather and share a meal that has been prepared and cooked together. In the same way that central heating in a modern home made the hearth less central and fragmented the time and space people shared, the microwave has enabled an atomization of cooking, eating, and the socializing that accompanies them.

Many of my informants, especially the reluctant habitual users, expressed a certain degree of regret at their reliance on their appliance and the ways in which their cooking and eating rituals have changed because of it. Yet they insisted that they do not have time to prepare meals any other way. Many avoiders cited the need to slow down as part of their motivation to stop using a microwave oven. In a process of self-indication (Vannini chapter five) they became conscious of the way this material object was altering the rhythm of their lives, and decided to negotiate their interaction with it. Savvy users, meanwhile, are comfortable and even enthusiastic about the ways their lives have been improved by the microwave. Like its technological ilk the cell phone and the laptop computer, the microwave oven causes many to marvel at the seemingly limitless possibilities of today and others to glance wistfully over their shoulders with nostalgia for the way we—Ding! Oops! I have to run! Food's ready!

References

Beaty, William. 2007. "Unwise Microwave Oven Experiments." Accessed September 23. Available at: http://amasci.com/weird/microwave/voltage1.html

Blumer, Herbert. 1972. "Symbolic Interaction: An Approach to Human Communication." Pp. 401-419 in *Approaches to Human Communication*, edited by R. Budd and B.D. Ruben. Rochelle Park, NJ: Spartan.

Bowden, Sue and Avner Offer. 1994. "Household Appliances and the Use of Time: The United States and Britain since the 1920s." *Economic History Review*, 47:725-748.

Cockburn, Cynthia and Susan Ormrod. 1993. *Gender and Technology in the Making*. London: Sage.

Cowan, Ruth Schwartz. 1983. *More Work for Mother: The Ironies of Household Technology from the Open Hearth to the Microwave*. New York: Basic Books.

Hammack, William. 2005. "The Greatest Discovery since Fire." *Invention and Technology*, 20. http://www.americanheritage.com/articles/magazine/it/2005/4/2005_4_48.shtml

Lemonnier, Pierre (Ed.). 2002. *Technological Choices: Transformation in Material Culture since the Neolithic*. London: Routledge.

Toys "R" Us. 2008. "Website Product Information: Kitchens and Play Food." Available at: http://www.toysrus.com/family/index.jsp?categoryId=2717639

U.S. Census Bureau. 2007. "The 2007 Statistical Abstract." Available at: http://www.census.gov/compendia/statab/construction_housing/housing_and_neighborhood_quality/ Last accessed April 3, 2008.

List of Contributors

Eugene Halton is Professor of Sociology at the University of Notre Dame. He has written countless journal articles and is the author of a forthcoming book, *The Great Brain Suck* (University of Chicago Press) as well as of *Bereft of Reason* (University of Chicago Press, 1995), *Meaning and Modernity* (University of Chicago Press, 1986), and coauthor with Mihaly Csikszentmihalyi of one of the great ethnographic classics in the field of material culture studies, *The Meaning of Things* (Cambridge University Press, 1981).

Ariane Hanemaayer is a Master's student in Sociology at the University of Waterloo, Canada. Her past ethnographic research focused on the sociology of sports and entertainment. Currently, she holds a research appointment with the Grey Zone of Health and Illness, an interdisciplinary Canadian Institute for Health Research (CHR)–funded project, and is doing ethnographic and phenomenological work on addiction and Alcoholics Anonymous.

Jon Hindmarsh is Reader in Work Practice and Technology in the Department of Management at King's College, London. He has written widely on the interactional constitution of the object, work practice, and video-based field methods. His recent publications include articles in *Organization Studies, The Sociological Review,* and *The Sociological Quarterly.* He coedited *Workplace Studies* (Cambridge University Press, 2000) and is currently coauthoring a text on video-based methods for Sage and coediting a collection of interactional studies of work practice for Cambridge University Press.

Grant Kien is an Assistant Professor at California State University, Eastbay, in the Department of Communications. Broadly stated, his research focuses on technography, qualitative approaches to technology research, globalization, communication and culture, wireless mobility, communications networks as performative, symbolic, and interpretive spaces, digital culture, new media audiences and consumption. His articles have appeared in *Qualitative Inquiry* and *Cultural Studies* ↔ *Critical Methodologies.*

Patrick Laviolette is Senior Lecturer and Director of Post-Graduate Studies in the School of Visual & Material Culture, Massey University. He has guest-edited a number of projects such as a series of articles on alternative sport for *Anthropology Today* (2007), a themed material culture issue for *Sites* entitled "Matter in Place" (2008), and a special issue on "anthropographic mappings" with Wystan Curnow for the *Journal of Visual Culture* (forthcoming). Recently he has contributed to *Reviews in Anthropology, City & Society,* and *The*

Senses & Society and is working on an intellectual biography on the early career of Sir Raymond Firth as well as preparing a coauthored book with Edward Relph entitled *Placelessness Revisited* (Ashgate, forthcoming).

Bryce Merrill is a doctoral candidate in the Department of Sociology at the University of Colorado-Boulder. His previous publications include expositions on Spencer Cahill's sociology of the person (in *Symbolic Interaction*) and interdisciplinary uses of narrative. His current research examines ethnographically the social psychological dimensions of home music recording, with a particular focus on home recording as an individually and socially consequential everyday practice.

Chaim Noy is an Independent Scholar whose research interests focus on semiotics, communication, and performance in everyday life. He teaches at the Hebrew University of Jerusalem, and at the Sapir and David Yellin colleges, in Israel. His recent research projects include ethnographies of driving and of mobilities in Jerusalem, Israel, and oral and inscriptional performances in tourism. Noy's recent book publications include *A Narrative Community* (Wayne State University Press, 2006), which deals with tourists' storytelling performances, and a coedited volume with Erik Cohen, titled *Israeli Backpackers* (SUNY Press, 2005). He has published articles and book chapters on reflexive and experimental methodologies (*FQS, International Journal of Social Research Methodology*), performance (*T&PQ*), tourism (*Annals*), gender, and sociolinguistics.

Tina Peterson is a doctoral student in mass media and communication at Temple University. Her research focuses on the impact of the media on dietary choices and food habits. As a former pastry chef, she is especially interested in how cooking skills and habits are learned and influenced.

Trevor Pinch is Professor of Science and Technology Studies, as well as Sociology, at Cornell University. One of the founders of the SCOT approach, Trevor Pinch is the author and editor of thirteen books: *Dr. Golem: How to Think about Medicine* (University of Chicago Press, 2005), *How Users Matter: The Co-construction of Users and Technologies* (MIT Press, 2003), *Analog Days: The Invention and Impact of the Moog Synthesizer* (Harvard University Press, 2002), *The Golem at Large: What You Should Know about Technology* (Cambridge University Press, 1998), *The Hard Sell: The Language and Lessons of Street-Wise Marketing* (HarperCollins, 1995), *Handbook of Science and Technology Studies* (Sage, 1994), *The Golem:*

What Everyone Should Know about Science (Cambridge University Press, 1993), *Dependency to Enterprise* (Routledge, 1991), *Health and Efficiency: A Sociology of Health Economics* (Open University Press, 1989), *The Uses of Experiment* (Cambridge University Press, 1989), *The Social Construction of Technological Systems: New Directions in the Sociology and History of Technology* (MIT Press, 1987), *Confronting Nature: The Sociology of Solar-Neutrino Detection* (Reidel, 1986), and *Frames of Meaning: The Social Construction of Extraordinary Science* (Routledge and Kegan Paul, 1982).

Ingrid Richardson is Senior Lecturer at Murdoch University, Western Australia, and has published on the cultural and corporeal effects of mobile media, virtual reality, biomedical imaging, technologies for sustainability, TV and public screens. Her research interests include philosophy of science and technology, new and interactive media theory, phenomenology, visual ethnography, and embodied interaction. Her publications have appeared in such journals as *Continuum, Fibreculture, Semiotica,* and the *Journal of Urban Technology.*

Mélanie Roustan, PhD in anthropology and sociology, University of Paris Descartes—Sorbonne, is an associated researcher to the CERLIS (Centre de recherche sur les liens sociaux, CNRS/Paris Descartes) and a freelance teacher (university of Paris Descartes, Sciences Po Paris). She published her thesis in 2007: *Sous l'emprise des objets? Culture matérielle et autonomie* (Ed. L'Harmattan); she is the coauthor of *Musée national des Arts africains et océaniens. Mémoires* (in cooperation with Anne Monjaret and Jacqueline Eidelman, and Bernard Plossu for the photographs, Ed. Marval, 2003) and she coordinated a collective book on video games, *La pratique du jeu vidéo: réalité ou virtualité?* in 2003 (Ed. L'Harmattan).

Amanda Third is Senior Lecturer and Director of the Centre for Everyday Life in the School of Media, Communication and Culture at Murdoch University, Western Australia. She has previously published in the areas of cultural studies, gender studies, and communication and media studies, and has recently completed a book-length manuscript on the popular cultural representation of female terrorists active in the 1960s and 1970s in the United States. Her publications have appeared in such journals as *Hecate, Social Alternatives, Animation Studies,* and *Theory & Event.*

Chris Tilley is Professor of Anthropology and Archaeology at the University College of London. Cofounder of the *Journal of Material Culture,* he is author and editor of thirteen books: *Domination and Resistance* (Unwin Hyman, 1989), *An Ethnography of the Neolithic: Early Prehistoric Societies in Southern Scandinavia* (Cambridge University Press, 1996), *Ideology, Power and Prehistory* (Cambridge University Press, 1984), *Interpretative Archaeology* (Berg, 1993), *Material Culture and Text: The Art of Ambiguity* (Routledge, 1991), *The Materiality of Stone: Explorations in Landscape Phenomenology* (Berg, 2004), *Metaphor and Material Culture* (Blackwell, 1999), *A Phenomenology of Landscape: Places, Paths, and Monuments* (Berg, 1994), *Post-glacial Communities in the Cambridge Region: Some Theoretical Approaches to Settlement and Subsistence* (B.A.R., 1979), *Re-constructing Archaeology: Theory and Practice* (Routledge, 1992), *Reading Material Culture: Structuralism, Hermeneutics, and Post-structuralism* (Blackwell, 1990), *Social Theory and Archaeology* (University of New Mexico Press, 1988), and the definitive guide to the field of contemporary material cultural studies, the *Handbook of Material Culture* (Sage, 2006).

Tanfer Emin Tunc is an Assistant Professor in the Department of American Culture and Literature at Hacettepe University. She earned her PhD from the State University of New York at Stony Brook in the social and cultural history of the modern United States, and specializes in women's history/literature; gender, sexuality, and reproduction; and feminist theory. She is the recipient of some of her field's most prestigious grants and fellowships (including awards from the National Science Foundation, the Woodrow Wilson Foundation of Princeton University, Rockefeller University, Duke University, the University of Michigan, Smith College, the American Historical Association, the National Women's Studies Association, the Society for the History of Technology, and the Francis C. Wood Institute for the History of Medicine at the College of Physicians of Philadelphia). Her most recent works include *Technologies of Choice: A History of Abortion Techniques in the United States, 1850–1980* (Verlag Dr. Muller 2008), *The Globetrotting Shopaholic: Consumer Spaces, Products, and their Cultural Places* (Cambridge Scholars Publishing 2008), and *Adapting America/America Adapted* (Edwin Mellen Press Forthcoming 2009).

Dylan Tutt is a Research Associate at Loughborough University's Department of Civil and Building Engineering. He specializes in visual ethnography and video-based studies of work practice and communication in diverse settings, including constructions sites, research labs, CCTV control centers, and the home. He has explored ways to apply these kinds of studies to the development and design of new systems and technologies. His recent publications confront "tactics" surrounding the use of household media (*Environment & Planning A*) and the domestic use of mobile phones by teenagers (*Convergence*), scrutinize interaction among distributed working teams (*ECSCW '07*), and encounter moments when the situated and mediated clash, or are crafted, into complex emotional encounters (*Qualitative Inquiry*).

Phillip Vannini is Associate Professor of Communication and Culture at Royal Roads University in Victoria, BC, Canada. He is coauthor of *Understanding Society through Popular Music* (Routledge, 2008) and coeditor of *Body/Embodiment: Symbolic Interactionism and the Sociology of the Body* (Ashgate, 2006). Phillip is currently carrying out a 4-year ethnographic project on the role that ferry boats play in the everyday life of island and coastal communities in British Columbia. Initial findings from that research have been published (or are currently in press) in the *Canadian Journal of Communication*, *Cultural Studies* ←→ *Critical Methodologies*, and in a forthcoming special issue of *Qualitative Inquiry* on technology and ethnography. Other studies of material culture and technology of his include ethnographic research on artificial tanning and the tanned body (published in *Symbolic Interaction* and *Qualitative Inquiry*), weather forecasting technologies (published in *Critical Discourse Studies*), music recording and distribution technology (published in *CTheory*), and the ritual of marriage proposal and the related exchange of romantic commodities (published in the *Journal of Popular Culture*).

Ian Woodward is a sociologist in the School of Arts, Griffith University, Australia and a Fellow of Yale University's Center for Cultural Sociology. His research on material culture, consumption, taste, and narrative is widely published in journals such as *The Sociological Review*, *Journal of Sociology*, *Journal of Contemporary Ethnography*, *Journal of Material Culture*, and *The British Journal of Sociology*. His book *Understanding Material Culture* has recently been published by Sage (2007). Woodward is also researching dimensions and practices of cultural openness and his research on cosmopolitanism with Zlatko Skrbis and Gavin Kendall has been published in *Theory, Culture and Society* and *The Sociological Review*. Their collaborative research in this area, "The Sociology of Cosmopolitanism," will soon appear as a monograph with Palgrave.

Index

A

Affordance 56; 60; 74; 75; 93; 101; 201

Agency 5-6; 20; 22; 32; 34; 38-41; 45; 76-78; 148; 150; 166; 171; 180; 182; 188; 196; 232; 239

Anthropocentrism ix-x

Autoethnography 101-103; 111-112; 195

B

Barthes, Roland 20-21; 64-65; 132

Body viii; 32; 38; 39; 41; 66; 78; 80-83; 89-99; 102; 116; 128; 145-147; 149-151; 153-155; 222

C

Choice 16-18; 20; 46; 48; 50; 56; 62; 98; 107; 112; 134; 162; 193; 194; 201; 242

Consumption x; 15; 18-20; 23; 28; 62; 64; 65; 99; 131-142; 151-152; 154; 159; 177-178; 180; 185; 193-194; 198; 207; 230-231; 237

D

Delegation 6; 32-33; 37; 53-54; 79; 80; 82

Domestication ix; xii; 17-18; 80; 104; 163; 193; 218; 234

Durability 27; 31-34; 38; 40; 178

E

Ethnography 4-6; 8; 31; 33; 38; 43; 78; 79; 83; 89-99; 121; 145; 147-148; 158; 160; 162; 167; 196; 202; 214; 232-233

Everyday life ix-xi; 6-8; 28-30; 42; 59; 64; 75-77; 81-82; 97-98; 101-106; 110-111; 115-116; 121; 128; 133-134; 137; 145-150; 153; 157-158; 167; 186; 189; 194; 196; 202; 207; 213; 216-217; 223; 224; 229

G

Gender 19-21; 69; 70; 78; 97-99; 108; 133; 148; 191; 214; 229

I

Identity x; 4; 18; 19; 21; 23; 36; 45; 61-62; 64; 67; 70; 80; 92; 108; 115; 118; 131-134; 136-142

Ideology 15; 20; 213; 224

Immutable mobile 27; 32; 38; 42; 212

Inscription 27-28; 33; 39; 41; 131

L

Latour, Bruno 2; 6; 27; 29-36; 40-43; 45-46; 51; 53-55; 131; 180; 197; 212

Lemonnier, Pierre 2; 17; 194; 197; 242

M

Materialism ix-xii; 15; 133
Materiality 1; 4; 15-16; 19-22; 28-30; 38; 40; 42; 45; 48; 54; 56; 59; 64-65; 70; 73; 78; 82; 91; 126; 131; 146; 148
Mattering 28; 42
Miller, Daniel 2; 15; 18; 22-23; 45; 93; 111; 138; 178
Mumford, Lewis ix; xii
Mutuality 17; 45; 51; 77; 102; 111;

N

Narrative 28; 35; 42; 59-70; 74; 92; 95; 195; 205; 223; 239; 242

O

Objectification 2-3; 6; 18; 21-23; 30; 59; 92; 99; 110; 182-183; 196; 206

P

Performance 8; 27-28; 30-33; 36-40; 42-43; 45; 52; 63-65; 73-74; 77-78; 91; 95; 99; 101; 104; 108; 109; 111; 131; 134; 141; 151; 159; 162; 166; 183-184; 193; 203-204; 206; 230; 232-234; 237; 241-242
Phenomenology 16; 21; 51; 110; 145-155; 173; 196; 205; 222
Post-structuralism 21; 81; 131
Power ix; 19-20; 30-32; 34-38; 41; 59-60; 65; 68; 74; 77; 80-83; 90; 92; 95; 97-98; 102-103; 108; 127; 137; 185; 213; 224

R

Relational materiality 27; 40
Ritual viii; 19; 20; 27-29; 63; 74; 77-78; 183; 234

S

Semiotics 15; 40; 42; 64; 65; 73; 74; 79-83; 131-133; 137; 140-142; 145-148
Skill x; 1; 3; 9; 32; 55; 62; 73; 81; 91; 93-94; 95; 153; 154; 186;
Space 30; 35; 39-40; 61-62; 65-66; 69-70; 74-76; 78; 91; 94; 96-97; 102-105; 108; 111; 131-132; 134-135; 141; 145-148; 150-151; 171-190; 193-195; 214-215; 220-223; 238; 240-242
Structuralism 21; 81; 131
Symbolic interactionism 36; 37; 73-83; 157-167; 229; 231-232

T

Techniques viii; xi; 4; 9; 16-18; 20; 31; 46; 48-49; 74-75; 82; 89; 91; 93-96; 131-132; 138; 140; 146; 155; 158-159; 161-166; 186; 196-197; 200; 205-206; 220; 223; 229; 234-235
Technics viii-xii; 4; 8; 17-18; 46-50; 53; 75; 78; 80;82; 131-139; 142
Techne 3; 6; 8; 23; 33; 82
Technoculture viii; x-xii; 4-6; 10; 15-16; 18-19; 23; 73; 75-79; 83; 157; 159; 207; 241
Time x; xii; 27-28; 30; 35; 38; 74-76; 91; 94; 96-99; 105-106;

115; 152; 178-179;181-182; 184; 187-188; 190; 198; 202-204; 212; 216; 218; 219; 222; 237; 239; 242-243
Transformation 17; 23; 41; 75; 77-80; 154; 194; 199
Translation 29-36; 38-42

U
Use 16-18

Intersections in Communications and Culture

Global Approaches and Transdisciplinary Perspectives

General Editors: Cameron McCarthy & Angharad N. Valdivia

An Institute of Communications Research, University of Illinois Commemorative Series

This series aims to publish a range of new critical scholarship that seeks to engage and transcend the disciplinary isolationism and genre confinement that now characterizes so much of contemporary research in communication studies and related fields. The editors are particularly interested in manuscripts that address the broad intersections, movement, and hybrid trajectories that currently define the encounters between human groups in modern institutions and societies and the way these dynamic intersections are coded and represented in contemporary popular cultural forms and in the organization of knowledge. Works that emphasize methodological nuance, texture and dialogue across traditions and disciplines (communications, feminist studies, area and ethnic studies, arts, humanities, sciences, education, philosophy, etc.) and that engage the dynamics of variation, diversity and discontinuity in the local and international settings are strongly encouraged.

LIST OF TOPICS

- Multidisciplinary Media Studies
- Cultural Studies
- Gender, Race, & Class
- Postcolonialism
- Globalization
- Diaspora Studies
- Border Studies
- Popular Culture
- Art & Representation
- Body Politics
- Governing Practices
- Histories of the Present
- Health (Policy) Studies
- Space and Identity
- (Im)migration
- Global Ethnographies
- Public Intellectuals
- World Music
- Virtual Identity Studies
- Queer Theory
- Critical Multiculturalism

Manuscripts should be sent to:

Cameron McCarthy OR Angharad N. Valdivia
Institute of Communications Research
University of Illinois at Urbana-Champaign
222B Armory Bldg., 555 E. Armory Avenue
Champaign, IL 61820

To order other books in this series, please contact our Customer Service Department:
(800) 770-LANG (within the U.S.)
(212) 647-7706 (outside the U.S.)
(212) 647-7707 FAX

Or browse online by series:
www.peterlang.com

www.ingramcontent.com/pod-product-compliance
Ingram Content Group UK Ltd.
Pitfield, Milton Keynes, MK11 3LW, UK
UKHW021829140426